INTELLIGENCE THROUGH SIMULATED EVOLUTION

WILEY SERIES ON INTELLIGENT SYSTEMS

Editors: James Albus, Alexander Meystel, and Lotfi A. Zadeh

INTELLIGENCE THROUGH SIMULATED EVOLUTION

Forty Years of Evolutionary Programming

LAWRENCE J. FOGEL
Natural Selection, Inc.
La Jolla, California

A Wiley-Interscience Publication
JOHN WILEY & SONS, INC.
New York / Chichester / Weinheim / Brisbane / Singapore / Toronto

Lyrics from *Experiment* by Cole Porter are used with the permission of Warner Bros. Publications U.S. Inc., Miami, FL. Copyright © 1933 (Renewed) by Warner Bros., Inc.

This text is printed on acid-free paper. ∞

Copyright © 1999 by John Wiley & Sons, Inc. All rights reserved.

Published simultaneously in Canada.

No part of this publication may be reproduced, stored in a retrieval system or transmitted in any form or by any means, electronic, mechanical, photocopying, recording, scanning or otherwise, except as permitted under Section 107 or 108 of the 1976 United States Copyright Act, without either the prior written permission of the Publisher, or authorization through payment of the appropriate per-copy fee to the Copyright Clearance Center, 222 Rosewood Drive, Danvers, MA 01923, (978) 750-8400, fax (978) 750-4744. Requests to the Publisher for permission should be addressed to the Permissions Department, John Wiley & Sons, Inc., 605 Third Avenue, New York, NY 10158-0012, (212) 850-6011, fax (212) 850-6008, E-Mail: PERMREQ @ WILEY.COM.

Library of Congress Cataloging-in-Publication Data:

Fogel, Lawrence J.
    Intelligence through simulated evolution : forty years of evolutionary programming / Lawrence J. Fogel.
      p. cm.
    "A Wiley-Interscience publication."
    Includes bibliographical references and index.
    ISBN 0-471-33250-X (alk. paper)
    1. Evolutionary programming (Computer science)   I. Title.
QA76.618.F64   1999
005.1—dc21   98-50703

Printed in the United States of America

10 9 8 7 6 5 4 3 2 1

To my son, David,
who showed me the great extent and utility of
evolutionary computation

CONTENTS

PREFACE ix

1 GENESIS 1

    1.1 Motivation / 3
    1.2 Prediction Experiments / 7
    1.3 Pattern Recognition and Classification / 20
    1.4 Control System Design / 28
    1.5 Extension of Early Evolutionary Programming Concepts / 31
    1.6 Competitive Goal-Seeking / 32
    1.7 Some Implications / 36
        1.7.1 Evolution and the Scientific Method / 36
        1.7.2 Evolutionary Problem Seeking / 40
        1.7.3 A Bound to Intellect / 40
        1.7.4 Evolution and Goals / 42
    1.8 Summary / 44

2 DIVERSIFICATION 46

    2.1 Two-Person Gaming against Nonminimax Players / 46
    2.2 Coevolution / 50
    2.3 Pursuit and Evasion / 52
    2.4 Modeling Time Series / 56

2.5 Pattern Recognition / 59
2.6 Simulated Ecosystems and the Nature of Intelligence / 65
2.7 Sequence Induction with Deterministic Automata / 65
2.8 Revisiting and Extending Early Evolutionary Programming / 68
2.9 Routing Problems / 75
2.10 Comparing Crossover, Inversion, and Mutation / 80

3 SPECIALIZATIONS 87

3.1 Finding Structure in Data / 87
3.2 Self-Adaptation / 96
3.3 Evolving Neural Networks / 111
3.4 Evolving S-Expressions and Multiple Interacting Programs / 119
3.5 Games / 122
   3.5.1 The Iterated Prisoner's Dilemma / 122
   3.5.2 Learning Tic-Tac-Toe without Knowing the Object of the Game / 127
   3.5.3 Evolving Strategies in Simulated Combat / 133
3.6 Other Applications / 135
3.7 Summary / 136

OUTLOOK 138

REFERENCES 143

INDEX 157

PREFACE

I received the Bachelor of Electrical Engineering degree from New York University in June 1948, the year the transistor was invented. To place this time frame in perspective, it was somewhere between the abacus and the laptop. I was thoroughly familiar with my Keuffel & Esser slide rule. All computation was rough-order analog—and that was good enough.

In 1948 it was hard to find a job in engineering. I finally had an opportunity to design antennas for various aircraft. This was a particular challenge in helicopter communication systems because of rotor modulation and the high-level noise produced by the engine. Even the very best antenna could not provide effective communication. I had to find a way to reduce the detrimental effect of noise. Why not separately sense that noise, invert that signal, then cancel the noise in both directions? This results in a very low ambient-level background for the pilot's reception of voice messages from the ground, and for the ground reception of his signal as well. The experimental demonstration was clumsy and unstable, but it worked. I received a patent on this device. Four decades later you can buy a variety of noise-canceling headsets. They are inexpensive, convenient, and effective. I later received other patents on other devices to improve communication and cockpit displays.

During the 1950s, I flew in many different aircraft. It occurred to me that the pilot's burden could be relieved by automating a part of his workload. I became interested in the way planes were being piloted and published a paper entitled "A Stochastic Model of the Pilot in Flight Control" in the November 1959 issue of the *Proceedings of the Institute of Radio Engineers*. Coincidentally, "What the Frog's Eye Tells the Frog's Brain" by Warren McCulloch and Jerry Lettvin appeared in the same issue. Evidently, we were trying to understand human information processing but from very different viewpoints.

In June 1960, I was asked to serve as assistant to the director of research at the National Science Foundation, the challenge being to determine how much America should invest in basic research. For example, what is the value of gaining new knowledge in each of the disciplines? Is it worthwhile to invest in a linear accelerator or Mohole Project? I focused particular attention on the prospect of artificial intelligence. There were two distinctly different schools of thought. The first engaged in "bionics," the attempt being to build an artificial brain, neuron by neuron. Leon Harmon's preliminary experiments at Princeton demonstrated the instability of artificial "neurons" operating in small chains. Frank Rosenblatt offered the Perceptron, but I was skeptical. How can an array of simple neurons replicate the function of the retina, no less the brain? Neural networks might be useful for pattern recognition, but intelligence is far more than that.

The second school of thought, termed "heuristic programming," was based on a psychological viewpoint. The idea was to achieve artificial intelligence through structuring generally useful if-then rules. Newell, Simon, and Shaw, and others, built programs that solved problems using this heuristic. Their approach to artificial intelligence is still being pursued, but it now is called knowledge-based or expert systems, as witness the work of Doug Lenat and many others.

But the fundamental questions are rarely asked—no less answered. How good an expert is required in this particular application? Is anyone qualified to serve as an expert? Over what domain are experts to be measured? How close is that domain to the current applications? When were the experts last scored? How well do they remember the rules they used? This last question is particularly important when experts must accomplish their task under significant pressure and/or in complex emotional situations. In such circumstances it becomes difficult to recall exactly what happened, and in what order, no less why. Heuristics are only *generally* useful rules. If a set of rules is truly appropriate for the current situation, it is likely to be far less appropriate at other times for the environment is dynamic and nonstationary.

Often there are a number of experts—so they are likely to disagree. If they each offer their best set of rules, how should these be combined? Should they be weighted by the level of expertise, length of experience, their currency, or what? In reality, experts don't just follow rules; they create them. They act to meet their view of the immediate need. As a result, they are less predictable and have a greater chance for success in a competitive environment. It is difficult to measure how apt each expert is in discovering new rules, and if we had such a measure, how would we use it to combine the expertise? Indeed, if intelligence is defined as the ability to solve *new* problems in *new* ways, expert systems are *unlikely* to lead to artificial intelligence.

Both bionics (neural networks) and heuristic programming (expert systems) attempt to model human behavior. But we are a mere artifact of the natural process of evolution, a process that generally yields creatures of increasing intellect over millennia. Why not model that process rather than this or any other particular artifact? I called this *evolutionary programming* (in contrast to heuristic programming), and presented the concept to my colleagues at the National Science Foundation in June

1961. Their response was totally negative. "You don't understand the immense number of possibilities. There are just too many alternative solutions. No procedure can search such a large domain in an efficient manner. It's like trying to find a needle in a haystack of haystacks. Trial and error will never get you there." I argued that natural evolution is trial and error *with inheritance* and *that's* what makes it an efficient learning process.

These objections continued for years. In 1966, Marvin Minsky denied the need for using a stochastic process. In his words, "Everything you have said in describing your technique in terms of evolutionary process ... is an old, but very suggestive biological idea. I submit that there has been technical progress in the area of artificial intelligence. Everything you have suggested ... can be put much more clearly and sharply in terms of tree searches."[1] And there were misunderstandings and misrepresentations of evolutionary programming by many other authors.

At this point it is important to understand that evolutionary programming was *not* intended to model natural evolution detail for detail, for that process is a singular occurrence—a sample size of one. Rather, the intent was to model evolution in more general form, that is, a process that occurs whenever reproductive entities exist in a finite noisy environment. By analogy, if you want to design an airplane, look to aerodynamics, structural mechanics, and the like. Do not try to replicate any particular bird. You do not need to put feathers on the wing to make it fly.

In July 1961, I returned to General Dynamics intent upon conducting research on evolutionary programming. I reasoned that an ability to predict one's environment is essential to intelligent behavior. How can one predict a nonstationary process with respect to an arbitrary payoff function? At that time Al Owens was head of the computer laboratory. He became interested in the experiments I had designed. Jack Walsh, a mathematician with the company, suggested that such experiments were unnecessary. There ought to be a way to build theorems and lemmas to determine the capabilities and limitations of this evolutionary process.

We formed a team. The first paper that presented our approach to artificial intelligence appeared in 1962. We had no knowledge of the work reported earlier by Friedberg and Fraser and had only heard of some experiments conducted by Bremermann. Our initial lack of interest in his work was partly because we heard that it had failed to demonstrate the worth of the evolutionary process—another unfortunate misrepresentation. After presenting several papers from 1962–1965, a book summarizing our experiments was published by John Wiley & Sons in 1966. *Artificial Intelligence through Simulated Evolution* was the first book in the field of evolutionary computation. Many other publications followed on our research and applications of evolutionary programming.

In 1965, Al Owens, Jack Walsh, and I formed a company, Decision Science Inc., to further explore the applications of evolutionary programming. It was the first company dedicated to making this technology a reality. I was president, Al was vice

---

[1] Comment reprinted in L. J. Fogel and W. S. McCulloch (1970) "Natural automata and prosthetic devices," *Aids to Biological Communication: Prosthesis and Synthesis,* vol. 2, D. M. Ramsey-Klee (ed.), Gordon and Breach, NY, pp. 221–262.

president, and Jack was secretary-treasurer. For 17 years, we developed and advanced evolutionary programming for the prediction of time series and modeling aircraft combat, and as an alternative to classic control theory. Many of the initial advances were achieved through the efforts of our friend and colleague George Burgin. The company grew, and in 1982 it was acquired by Titan Systems, Inc., a defense contractor in San Diego, California that later went public.

We didn't publish much between 1968 and 1982. In those years we felt that it was better to hold our ideas "close to the vest," since we offered a unique technology, a brand new way of solving problems. We didn't want to create competitors. Only a few people were still exploring evolutionary computation in the 1970s, and most of them were at universities. Their rule was publish or perish, but we viewed the situation differently: publish *and* perish.

In 1984, my son David began working part-time at Titan Systems, Inc. while completing his undergraduate studies at the University of California at Santa Barbara. I had just won a contract to explore some basic research questions in evolutionary programming from the Army Research Institute. For a year, together we went back to my early efforts in the early 1960s and recapitulated them on faster machines, exploring aspects of evolutionary programming for which I could only offer questions 25 years earlier. David extended my early ideas by applying evolutionary programming not just to the adaptation of finite state machines but to optimizing the coefficients of polynomials for modeling data, finding optimal sequences for traveling salesmen problems, designing neural networks, and in many other areas. It was a time of renaissance to see the original ideas flower in so many different directions.

I was also fortunate to have Wirt Atmar serve as a consultant in my work with David. Wirt received the ScD degree, actually a double doctorate in biology and electrical engineering, from New Mexico State University in 1976. His dissertation was on evolutionary programming, under the direction of Don Dearholt, whom I later found out had supervised many other master's students in evolutionary programming during the 1970s. Wirt served as our source on biology. He showed us what was correct, what was incorrect, and what we just didn't know. He also spent hours mentoring David, giving freely much of his time. With David's enthusiasm and Wirt's guidance, evolutionary programming was given new life.

David starting publishing our work, and later his own. Coincidentally, interest in simulating evolution began growing. The first conference on "genetic algorithms" was held in 1985, and the conferences kept getting larger in 1987 and 1989. I later learned of a group in Germany that was pursuing similar methods of evolutionary computation at the University of Dortmund and other locations. They had started a conference series in 1990. Evolutionary computation was on its way to becoming legitimate in the scientific community.

In preparing his dissertation in engineering at the University of California at San Diego, David convinced his advisor Tony Sebald that evolution could be used to train neural networks instead of backpropagation. Some of his peers at school also became interested, notably John McDonnell. By this time, we had both moved over to a new company, ORINCON Corporation, and several of our coworkers were interested in exploring the use of evolutionary algorithms for sonar signal recognition problems. By 1992, we had a sufficiently large community to hold the First Annual Conference on Evolutionary Programming. The conference has grown in size every year since, and there are now hundreds of papers on evolutionary programming that have been published in conference proceedings and journals internationally.

It has been almost 40 years since I laid out this fundamentally new approach to artificial intelligence. At the very least the idea is now sufficiently well accepted so as to be controversial. In the 1960s computers were too slow to be capable of achieving practical results on complex problems. Today desktop machines are all that is necessary. No doubt the speedup of computer technology ushered in the amazing explosion of interest in evolutionary programming, and more broadly in evolutionary computation as a whole. It makes this opportunity to revisit the efforts in evolutionary programming in detail all the more timely.

This book is the second edition of my seminal 1966 work with Owens and Walsh. Chapter 1 essentially recaps that earlier book; Chapters 2 and 3 discuss the work in evolutionary programming from the 1970s to the present day. As we come to the end of the millennium, we come to the end of four decades of research in evolutionary programming. This book gives us an opportunity to extend our efforts to evolutionary computation in general. I hope that the many applications of evolutionary programming that this book offers will serve as a foundation to move the artificial intelligence community forward.

There are too many people to thank for their help over the years. I regret that I cannot name them all. Wirt Atmar, Julian Bigelow, Warren McCulloch, and my wife, Eva, provided inspiration. Pete Angeline, George Burgin, Kumar Chellapilla, David Fogel, Gary Fogel, Jong-Hwan Kim, John McDonnell, Brian Moon, Al Owens, Ward Page, Bill Porto, Bob Reynolds, Tony Sebald, Jack Walsh, and Xin Yao contributed to developing the concept of evolutionary programming and specific applications. Jim Albus, Tom Brotherton, Art Callahan, Michael Conrad, Don Dearholt, Howard Dunholter, Steve Freer, Myles Maxfield, Alex Meystel, Capt. Dennis McBride, Gene Ray, Terry Rickard, Matt Rizki, and Lotfi Zadeh were most supportive, as have been the members of the IEEE Neural Networks Council.

Thanks so much.

LAWRENCE J. FOGEL

La Jolla, CA

CHAPTER 1

GENESIS

Life on earth has evolved for some 3.5 billion years. Initially, only the strongest creatures survived the adversity of the natural environment and the competition of other living organisms. As time passed, certain species grew sufficiently complex to recall the repetition of similar events. As a result these creatures developed skills, and soon the rule became "survival of the most skillful." That individual survived which could consistently sense danger, evade threat, or utilize camouflage. The very existence of humans stands as testimony to the ability of our forebears to outwit, rather than outpower, in the competition for survival. Clearly, the ability to survive was measured by how consistently an individual made the proper choice from among available alternatives.

With increased sensory capability, it became possible for creatures to recognize the existence of new situations that required decisions outside the realm of "conditioned reflex." The individual had to draw on whatever analogies could be found between the observed situation and those in which his repertoire of available skills had previously demonstrated its worth. This was the beginning of what might be called intelligent behavior. Under the rigors of continual struggle, it was often the more intelligent creature that survived to reproduce more of its kind. Some species "solved" the problem of survival by having an increased number of offspring, whereas others decreased the scope of their activity in order to exclude the apparent threat. Although these are of interest, particular attention will be devoted to "creatures" whose response to the threat of their environment is through intellectual adaptation.

This ability to discover aspects of similarity also provides a capability for self-observation and self-appraisal. Interest in making wise decisions was a natural outgrowth. Only in modern times has this interest been accompanied by an inquiry into the very nature of the decision-making process itself. The growth of technology has provided us with new tools that may make it possible to synthesize automata which display features characteristic of our decision-making behavior. The modeling of humans may result in mechanisms that replicate this capability but perform it with much greater speed, accuracy, and consistency.

But it is of far greater interest to search for an understanding of the logical properties that constitute intellect. One means toward this end is the study of the natural processes by which creatures of greater intelligence have evolved. Natural evolution is recognized to be an iterative process consisting of mutative reproduction, natural selection, voluntary recombination, individual learning, and so forth. It is hoped that through a replication of specific aspects of evolution, means will be found for the generation of artificially intelligent automata, which are capable of solving problems in previously undiscovered ways. This may well provide a deeper understanding of the very organization of intellect.

Interest in intelligent decision making has existed for a long time. Effort was first devoted to aiding the decision maker with sound advice and relevant information. Decision making was treated as an art. Not until the late nineteenth century was any serious inquiry made into the behavioral aspects of intelligent decision making. As Lord Kelvin (W. Thomson) remarked:

> I often say that when you can measure what you are speaking about, and express it in numbers, you know something about it; but when you cannot measure it, when you cannot express it in numbers, your knowledge is a meagre and unsatisfactory kind; it may be the beginning of knowledge, but you have scarcely, in your thoughts, advanced to the stage of science, whatever the matter may be.

The distinction between knowledge and intelligence became clear; knowledge was useful information stored within the individual, and intelligence was the ability of the individual to utilize this stored information in some worthwhile (goal-directed) manner.

As effort to measure the intelligence of decision making has progressed, there has been a concurrent inquiry into the logical structure of intellect. This inquiry has grown out of purely philosophical speculation into the mechanization of various hypotheses with a view toward demonstrating their validity. It is now generally agreed that a better understanding of the organization of intellect can be demonstrated by constructing a device that exhibits what is said to be "intelligent behavior." The last two decades have seen significant surge of interest in such mechanizations.

Much of this interest has been directed along two lines of thought. The first recognizes the human animal to be the most intelligent creature in nature. It would appear worthwhile, therefore, to replicate human intellect in some specific manner such as in terms of our "neural networks." The second approach agrees that humans are most worthy of modeling but prefers to view humans as a psychological entity. Such modeling may be accomplished by presenting situations to a human subject who is required to offer a response or "move." Their decision as to which move to make is then analyzed in order to reveal some consistent set of subquestions that facilitates this decision making. Consistent aspects of this rationale become the basis of a computer program, which is hoped to "play the game" in a similar manner. This approach was first called "heuristic programming." It later grew into knowledge-based or expert systems.

Neither of these approaches has succeeded in making the inroads into the fundamental nature of intellect that were once promised. Models of neural networks have greater potential in this regard, although the sheer size of the human brain and its connectivity pose significant challenges to its complete reverse engineering. In contrast, expert systems are fundamentally limited to address questions for which we already know the answers. In the absence of an expert, there is no "system."

A third approach is possible. From a less egocentric standpoint, the human animal may be viewed as a single artifact of the natural experiment called evolution. Though, certainly, man is an intelligent creature, there is no reason to believe that we are the most intelligent creature that could possibly exist. Nature's experiment is continuing, and, in view of its past success, it would appear quite reasonable to anticipate that some future creatures will possess far greater intelligence than present-day humans. In this light, it would appear worthwhile to replace the process of modeling man as we now exist with the process of modeling evolution. This evolutionary process would be carried out in order to reach a better understanding of the main properties of intellect. By this insight, we hope to gain a means for the synthesis of machines that display greater intelligence than has thus far been found in nature. In contrast to heuristic programming, we propose a method of *evolutionary programming* as an alternative method for generating artificial intelligence.

1.1 MOTIVATION

Intelligent behavior is a composite ability to predict one's environment coupled with a translation of each prediction into a suitable response in light of some objective (Fogel et al., 1966, p. 11). Success in predicting an environment is a prerequisite to intelligent behavior. For simplicity, but without sacrificing generality, Fogel et al. (1966) considered the environment to be a sequence of symbols taken from a finite alphabet. The task was to devise an algorithm, essentially a computer program, that would operate on the observed indexed set of symbols and produce an output symbol that is likely to agree with the next symbol to emerge from the environment.[1] Evolutionary programming offered a means toward this end, and the basic procedure adopted was as follows.

A computer program was written to conduct fast-time simulation of some of the fundamental aspects of natural evolution. A collection of alternative algorithms, often chosen at random over some appropriate sample space, formed the initial population. These algorithms were evaluated based on how well they predicted each next symbol in an initial known sequence by operating on previous symbols. The algorithms were then randomly altered, or "mutated," into new forms, with the offspring

---

[1]Other payoff functions could also be devised. For example, the process might just as easily be made to generate algorithms that would disagree with the next symbol. Any arbitrary payoff function could be used.

evaluated by the same predictive task and scored in a similar manner as their parents. The best-scoring subset of some chosen size from the parents and offspring was selected to be retained as parents for the next iteration. Algorithms that scored below the threshold were removed. And the process was continued through successive rounds of mutation and selection until an algorithm of sufficient credibility was discovered or the available time had elapsed.

With consideration being given to environments defined by sequences of symbols from a finite alphabet, finite state (Mealy) machines provided a convenient mechanism for representing predictive algorithms. Specifically, each finite state machine possessed a finite number of states and could respond to a finite number of input symbols. For each possible input symbol, in each state, the machine would generate an output symbol and follow a state transition (possibly to remain in the same state). For example, Figure 1.1 shows a three-state machine. The states are labeled A, B, and C. Presume that the machine is in state A. If a "1" is sensed, the machine will respond with a "β" and remain in state A; however, if a "0" is observed, the machine will again output a "β" but made a transition to state B. Note that the input alphabet and output alphabet do not need to be the same, although they were the same in the early experiments in evolutionary programming.

Consider the problem of evaluating the quality of a machine used to predict a cyclic environment (101110011101)*, where the * indicates that the sequence inside the parentheses cycles indefinitely. Let us presume that the first four symbols from this sequence are available, and we will evaluate the machine shown in Figure 1.2 over the next eight symbols. As shown in the figure, the machine responds to the environment by generating a sequence of symbols where four of the eight predictions are correct. The machine in Figure 1.3, in contrast, predicts each next symbol

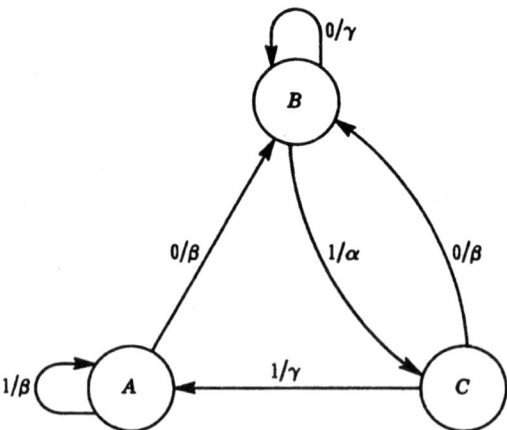

Figure 1.1 A three-state finite state machine. The machine is presumed to start in state A unless otherwise stated. Input symbols {0, 1} are shown to the left of the virgule, output symbols {α, β, γ} are shown to the right of the virgule. The arrows indicate the next-state transition for each input symbol.

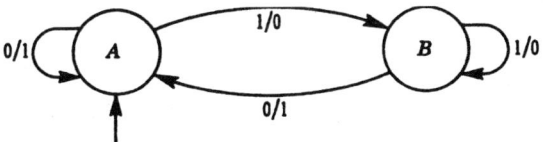

```
Environment:    1  0  1  1  1  0  0  1  1  1  0  1
FSM Predicts:         0  1  0  0  0  1  1  0  0  0  1
Current State:        A  B  A  B  B  B  A  A  B  B  B  A
```

Figure 1.2 The two-state machine generates the sequence 00110001 for the 5th through 12th symbols in the sequence (shown in bold). This machine is correct in four of these eight predictions for an average worth of 0.5. The initial observed environment is assumed here to be four symbols, but this choice is arbitrary.

in the sequence perfectly. In order to measure the quality of various machines, a suitable payoff function must be defined that establishes a value for every possible correct and incorrect prediction. The simplest might be the all or nothing case: If the machine predicts the next symbol correctly it receives a point; otherwise, it receives no payoff. Over a series of predictions, a machine's effectiveness might be reasonably taken to be the average value of its predictions.

In the original evolutionary programming experiments, new machines were created by making random variations to existing machines. The state-transition diagram suggests five obvious ways of varying a finite state machine. The first is to add a state and then randomly assign all the input-output and input-transition pairs for this state. This can be accomplished provided that the maximum machine size is

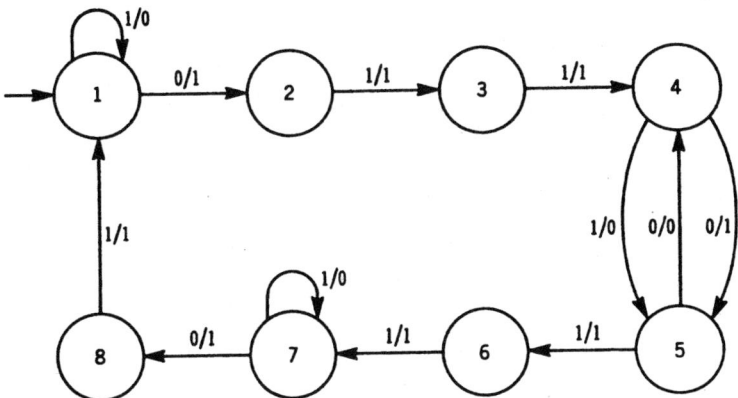

Figure 1.3 The depicted eight-state machine can generate the sequence (101110011101)* perfectly. Note that not all of the states are defined for every possible input symbol. The omitted symbols are left out because they are not necessary in defining the behavior of the machine to the specified input sequence (from Fogel et al., 1966, p. 31).

not violated due to memory constraints. The second is to delete a state if there is more than one state in the machine. Any state transitions that point to this state must then be redirected to other existing states, typically chosen uniformly at random. The third is to change an output symbol coupled with a specific input symbol in a single state. The fourth is to alter a state transition associated with an input symbol in a single state. Finally, the fifth is to change the starting state if there is more than one state. By altering the probability of each of these mutations occurring, or even the number of mutations introduced, a significant diversity of new machines can be created from existing parents.

In the earliest efforts in evolutionary programming, selection was applied so as to only retain the best available finite state machines. For convenience, the population size was kept constant from iteration to iteration by limiting it to three or five machines. The small population size was a computational necessity given the available computing hardware. Using current notation, each of μ parents was mutated to create a single offspring, with the 2μ solutions competing for survival. The best μ were retained as parents for successive progeny. The possibility for including less fit offspring was considered (Fogel, 1964; Fogel et al., 1966, p. 21), but due to computing speeds such practice was typically avoided.

The basic procedure is shown in Figure 1.4. Several iterations of mutation and

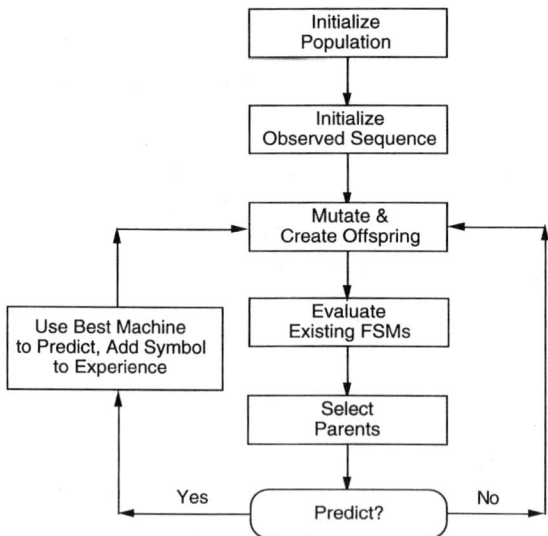

Figure 1.4 Flowchart of early evolutionary programming. A population of finite state machines is initialized at random within memory limitations on the maximum number of states. The known sequence of symbols from the environment is defined. The existing parents are mutated to yield offspring machines. All of these parents and offspring are evaluated on how well they predict each next symbol in the known sequence of symbols. The best half of these finite state machines are selected to become parents for the next iteration. If required, the best available machine is used to predict the next symbol, and the actual symbol is added to the known available data. The process is iterated for the time allowed.

Table 1.1 State-transition and output symbol matrix for the initial five-state machine used in the original evolutionary programming experiments summarized in Fogel et al. (1966)

State	Input Symbol	Next State	Output Symbol
1	1	2	1
1	0	1	1
2	1	3	1
2	0	2	0
3	1	4	1
3	0	5	0
4	1	5	1
4	0	3	0
5	1	1	1
5	0	2	0

selection were conducted before requiring a prediction of the next symbol in the environment. The best available machine served as the basis for that prediction, and once the next symbol was observed, it was included in the known environment and the process was continued.

This approach, first offered in the literature in Fogel (1962a), was studied in a series of experiments published in Fogel (1964), Fogel et al. (1964, 1965a, 1965b, 1965c), and others, and summarized in Fogel et al. (1966). Some examples from these early experiments are provided below.

1.2 PREDICTION EXPERIMENTS

A series of increasingly difficult prediction experiments was conducted in order to demonstrate the feasibility of evolutionary programming. Attention was first given to a two-symbol environment (i.e., the symbols were either 0 or 1), with the evolutionary program being written in Fortran II on an IBM 7094.[2] The five-state machine shown in Table 1.1 was used to seed the initial population (all individuals in the population being set to this same machine). At each point in time, the best three machines were retained to serve as a basis for generating offspring.[3] Five iterations (originally termed *generations*) of mutation and selection were performed before the best available machine was used to generate a prediction of the next symbol in the environment, whereupon this symbol was included in the observed sequence and the process continued.

---

[2] For the sake of comparison, David Fogel contacted IBM in the late 1980s and learned that this machine had essentially half the speed of an Apple II or Commodore 64.
[3] Although a population of solutions was used in all of the early evolutionary programming experiments, the method was occasionally incorrectly described as relying on only a single parent (e.g., Holland, 1975, p. 163).

The first series of experiments were aimed at predicting an environment generated by an unending repetition of the arbitrary sequence 101110011101. The first four experiments examined the sensitivity of the procedure's capability to predict symbols in the sequence as a function of the types of mutation that were imposed on parent machines. These initial experiments did not impose a penalty for complexity of the evolving finite state machines, and only a single mutation was applied to each surviving parent in order to generate each offspring machine. The mutation noise was uniformly distributed over the five modes of mutation: (1) add a state, (2) delete a state, (3) randomly change a next state, (4) randomly change the initial state, and (5) change the initial state to the second state the machine assumed under the available experience. The last of these mutations was used to permit the possibility of "phase shifting." The sequence of random numbers used for each experiment was drawn from *A Million Random Digits with 100,000 Nominal Deviates*, published by Rand Corporation, Free Press of Glencoe, New York, 1955.

Figure 1.5 shows the results from four experiments in terms of the percent correct as a function of the number of symbols experienced from the environment. Several thousand offspring were evaluated in each of these experiments, and the eventual predictor machines often grew to around eight to ten states. In experiment 4, a sequence of perfect predictor-machines was found after the 19th symbol of experience. Poorest prediction occurred in experiment 3, but even this experiment demon-

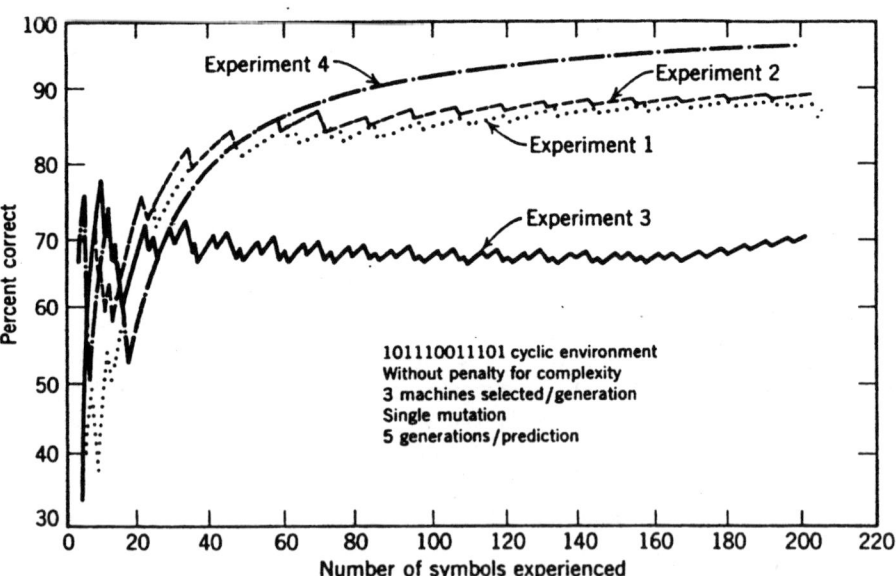

Figure 1.5 A comparison of the cumulative prediction score for four different random trials of evolutionary programming predicting the sequence (101110011101)*. No penalty for complexity was imposed on the finite state machines. Three parents were kept at each iteration, with offspring being created by a single instance of mutation to each parent (from Fogel et al., 1966, p. 28).

strated a distinct tendency for the quality of predictions to improve after the first dozen repetitions of the basic pattern of the cyclic environment. The first experiment was considered typical and will be used as a basis for comparison in what follows here.

The effect of imposing a penalty for machine complexity is shown in Figure 1.6. The solid curve is the result of experiment 5, which was identical to experiment 1 except that a penalty of 0.01 per state was imposed. Note that a significant difference in prediction capability was apparent only in the initial period. The benefit derived from the use of such a penalty for complexity is evident in Figure 1.7, which shows the significant reduction in the complexity of the evolved predictor-machines. It is interesting to note that perfect prediction of this repetitive environment (101110011101)* can be achieved by various eight-state machines (three of which are shown in Fig. 1.8). Restricting the number of states may preclude finding any of these machines.

With an evident need for larger predictor-machines to treat more complex environments, it is reasonable to suspect that an increase in the probability of adding a state might improve the prediction capability. Thus experiment 6 was a repetition of experiment 1 except that the probability of adding a state was increased from 0.2 to 0.3 (this increase being balanced by a reduction in the probability of deleting a state). Figure 1.9 shows the benefit derived. A sequence of perfect predictor-machines was found after the 19th symbol experienced.

Another way to obtain larger predictor-machines is through the use of multiple

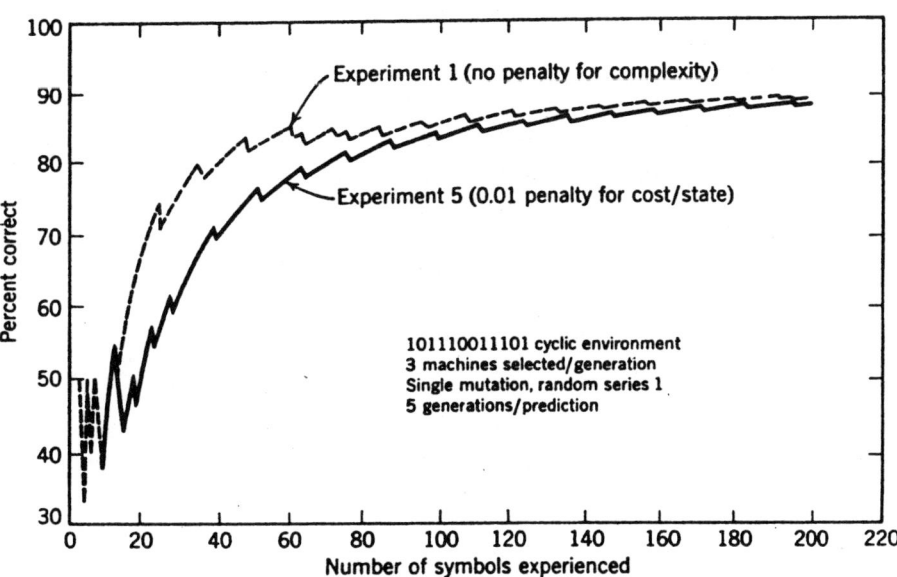

Figure 1.6 A comparison in two experiments of the prediction score with and without a penalty for complexity. The penalty imposed in experiment 5 was 0.01 per state (from Fogel et al., 1966, p. 29).

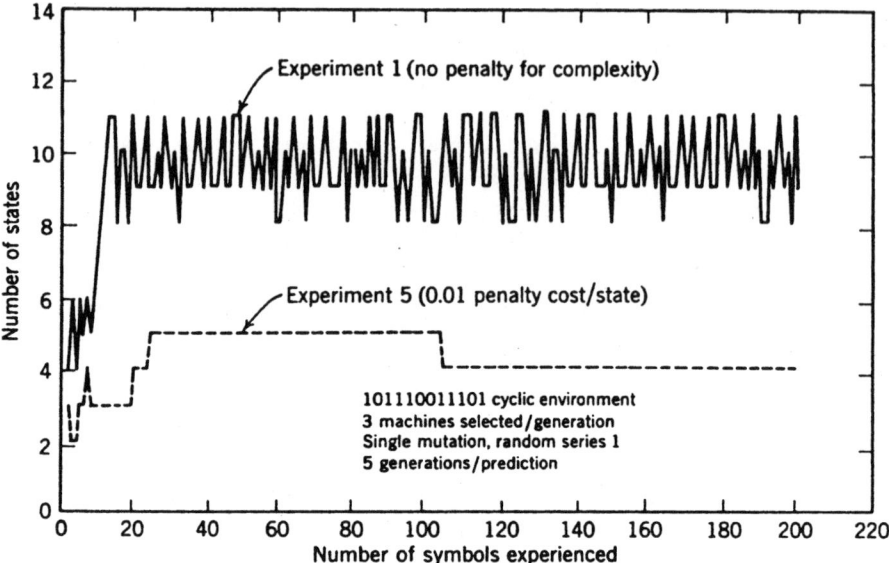

Figure 1.7 A comparison in the two experiments depicted in Figure 1.6 of the number of states of the best predictor machines with and without a penalty for complexity of 0.01 per state. The size of the evolved machines reflects the inclusion or absence of the penalty (from Fogel et al., 1966, p. 30).

mutations per parent. Figure 1.10 shows the results of experiments 1, 7, and 8, involving the use of one, two, and three mutation operations applied to each surviving parent machine, respectively. As might be expected, there was an increase in the predictive score with an increase in the multiplicity of mutation, since this allows for essentially a larger possible step size of variation. The size of the predictor-machines for each of these experiments is shown in Figure 1.11.

It was interesting to consider the effect of the length of initial recall (in terms of the number of symbols) on the prediction capability. For a purely cyclic environment in the absence of any noise, increasing the length of recall provides for a larger sample size with no possibility for overfitting the observed data; however, in the case of a nonstationary and/or noisy environment, it may be better to forget past events after some amount of time. Figure 1.12 shows a comparison of the prediction score for experiments with initial recall lengths of 2 symbols and 20 symbols (experiments 9 and 10, respectively). During the initial sequence experience, the predictive behavior appears nearly random, but the longer recall did exhibit faster learning of the cyclic environment.

It was also of interest to predict environments that undergo a sudden radical change in behavior. Figure 1.13 shows the result of experiment 11 in which the environment consisted of the sequence 101110011101 repeated ten times; this was followed by the inverse sequence 010001100010 repeated ten times. The initial recall was 20 symbols long; that is, the 21st symbol to emerge from the environment

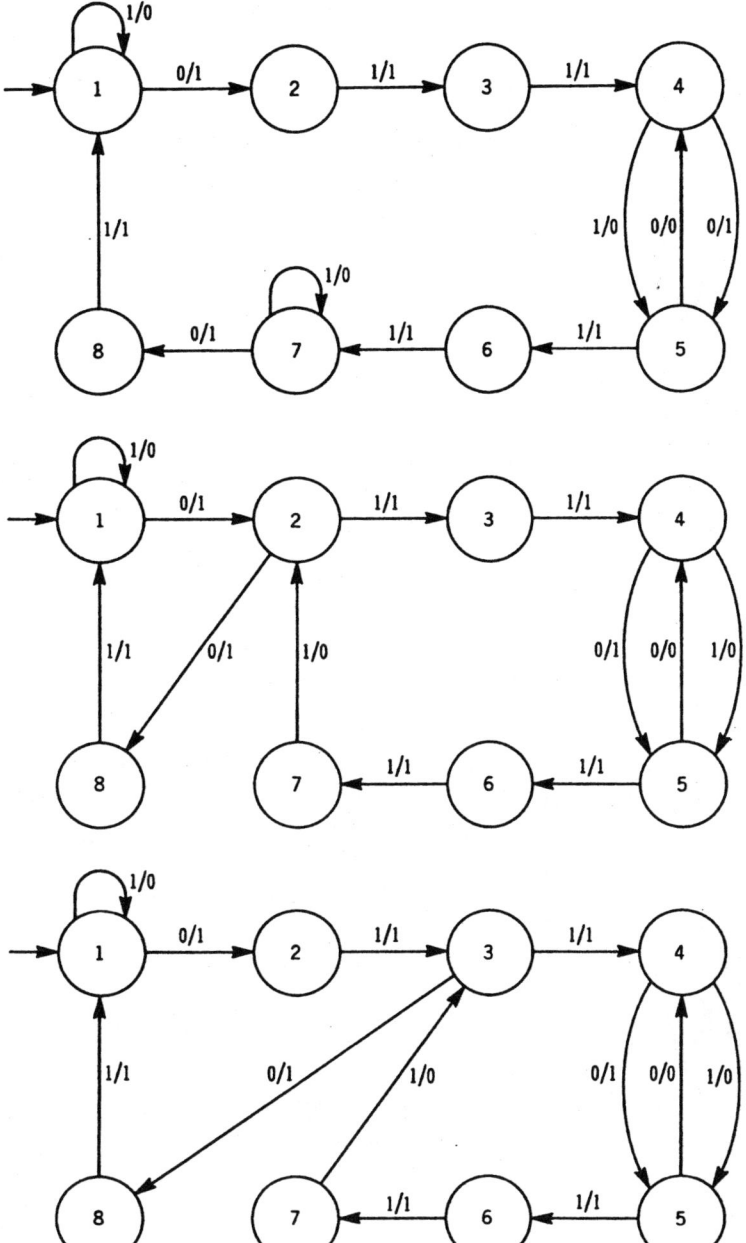

Figure 1.8 Three different eight-state machines that predict the sequence (101110011101)* perfectly (from Fogel et al., 1966, p. 31).

12 GENESIS

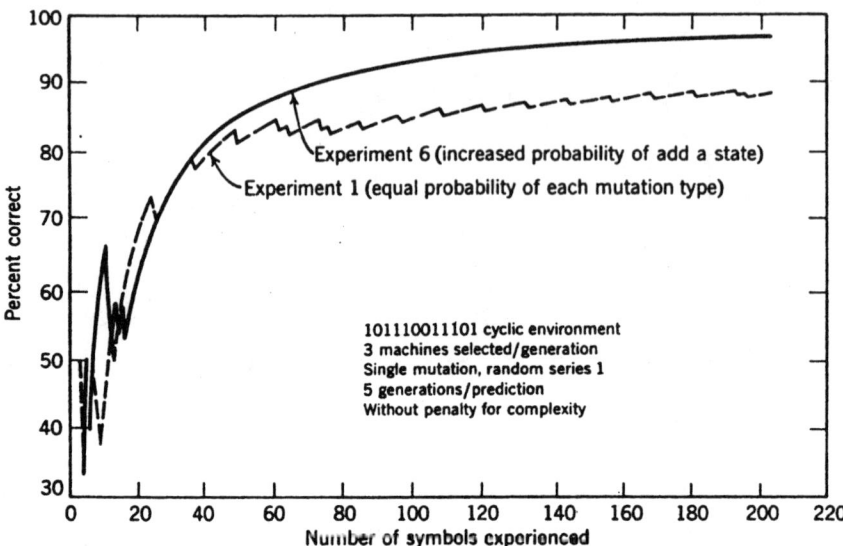

Figure 1.9 The result of increasing the probability of adding a state to 0.3. The increased probability of growing the size of the evolving machine translated into an improved ability to predict the given environment in this trial. A perfect machine was found after the 19th symbol experienced (from Fogel et al., 1966, p. 32).

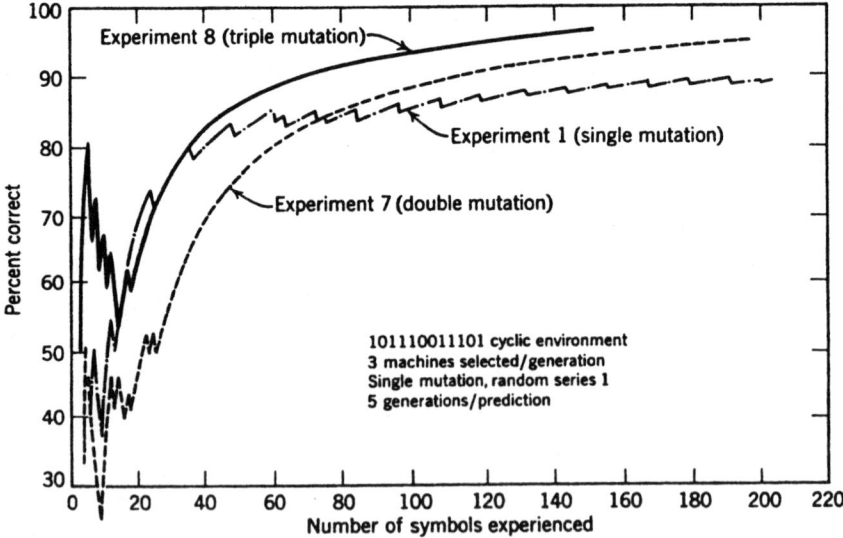

Figure 1.10 Increasing the multiplicity of mutation can also lead to increased probability of added states. The figure indicates a comparison of trials with one, two, and three mutations per parent in creating each offspring. The two trials will multiple mutations discover perfect predictors (from Fogel et al., 1966, p. 32).

1.2 PREDICTION EXPERIMENTS   **13**

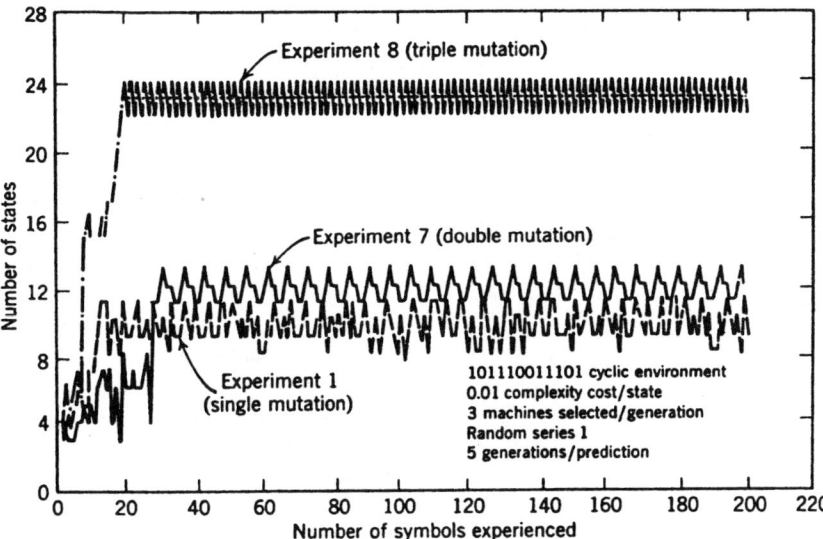

Figure 1.11 The size of the evolved predictor machines shown in Figure 1.10. The trials with multiple mutations per parent generated larger machines, which were more capable of generalizing the experienced environment (from Fogel et al., 1966, p. 33).

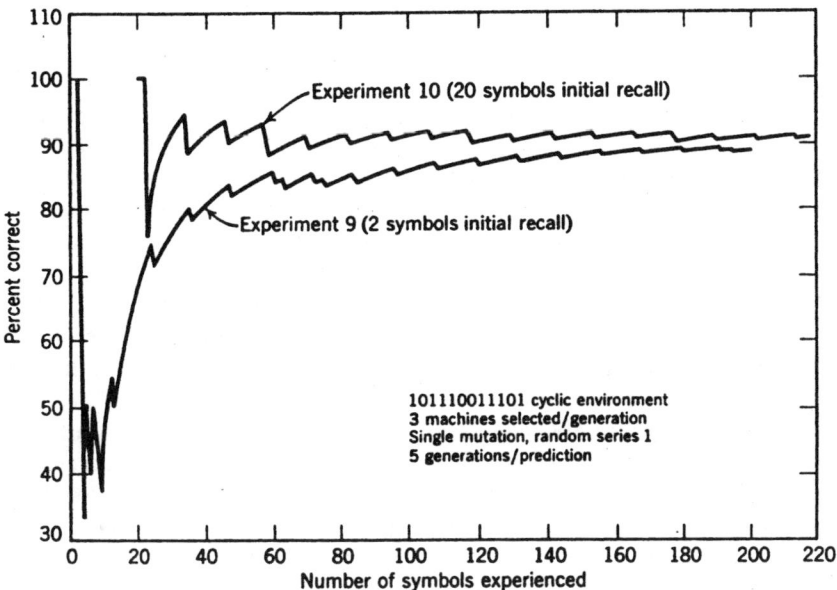

Figure 1.12 Experiments varying the symbol recall length. Allowing for a longer recall, and thus a greater memory, improved the initial learning rate (from Fogel et al., 1966, p. 34).

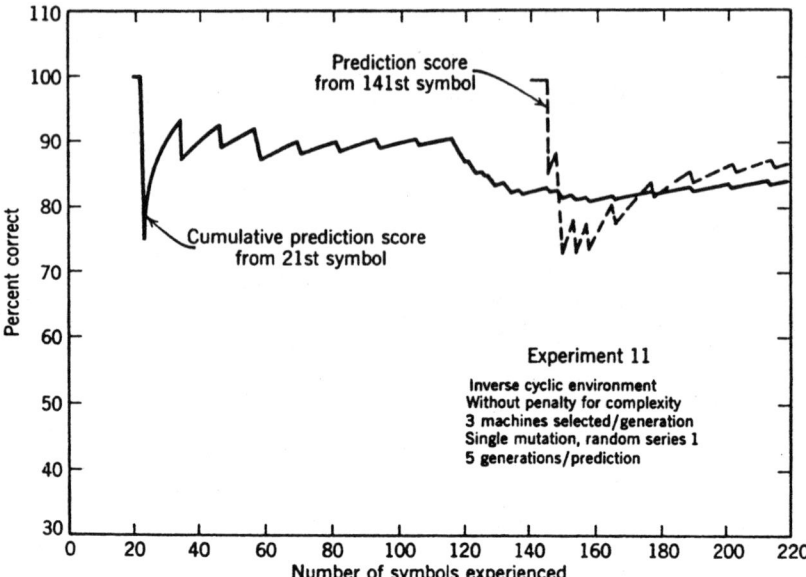

Figure 1.13 An experiment involving a radical change in the environment. At the 120th symbol, the environment switched from (101110011101) to (010001100010). The performance of the population of evolving machines degrades until sufficient experience with the new environment is attained (from Fogel et al., 1966, p. 34).

was the first symbol predicted. This initial recall permitted the selection of machines that correctly predicted the 21st, 22nd, and 23rd symbols before an error was made (when the environment changed). The prediction score then fluctuated around 0.9 until the first phase of the environment came to an end. At this point a series of errors occurred, and the prediction score dropped to 0.814 by the 159th symbol. Thereafter the selected machines exhibited "adaptation," and the prediction score again began to rise.

To portray this adaptation, it is fitting to replot the prediction score, beginning with the 141st symbol. The dashed curve in Figure 1.13 indicates the prediction score for the second phase of the environment, but with the initial machines resulting from that phase of experience. It is evident that some "unlearning" had to take place. This second-phase prediction score compares favorably with that of the first phase when it is taken into account that the machines are measured in their predictive fit over the entire length of experience. This explains the rapid growth in the number of states of the machines used for prediction, as shown in Figure 1.14: The evolving machines became more complex to handle the shift in the observed environment. During the first phase of the environment the number of states had remained near 6, whereas in the second phase the number of states stabilized in the vicinity of 16.

A second level of difficulty was imposed by using a statistically stationary envi-

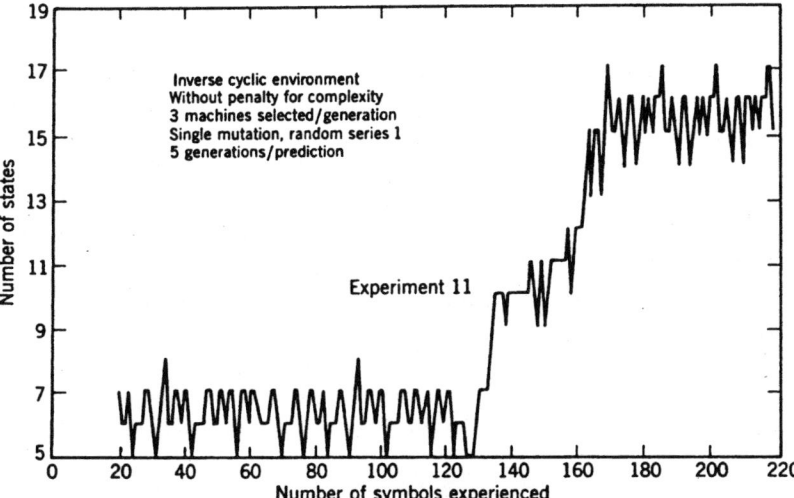

Figure 1.14 The number of states of the best evolving predictor machine as a function of the number of generations. When the environment shifts from (101110011101) to (010001100010) the size of the predictors increases, retaining the ability to fit the past history of the first environment while gaining sufficient flexibility to predict the new environment (from Fogel et al., 1966, p. 35).

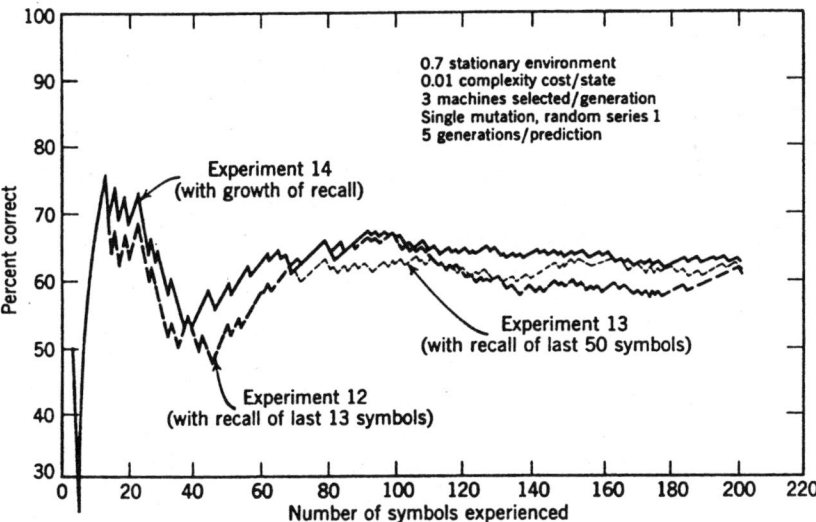

Figure 1.15 Three experiments involving a stationary environment where the probability of a 1 is 0.7. A variety of recall lengths (effective sample size) indicated a tendency for improved initial learning when using a longer recall period (from Fogel et al., 1966, p. 36).

ronment of 0's and 1's that resulted from the repetitive flipping of a 0.7 weighted coin. With a priori knowledge of this fact, it would be possible to predict this environment with an asymptotic score of 0.7, given an all-or-none payoff function. Figure 1.15 indicates the result of applying the evolutionary procedure with no such prior knowledge. The effect of three different lengths of recall are shown. Restriction of the recall to the last 13 symbols (experiment 12) reduced the initial prediction capability significantly (this length of recall was chosen to correspond with the length of experience upon reaching the first peak of average prediction score under continually growing recall). In contrast, recalling the last 50 symbols (experiment 13) produced only slightly poorer results than no restriction in the length of recall. However, this constraint improved the efficiency of the computation process. It greatly reduced the computation time required for evaluation of each machine, especially in longer sequences of experience. Experiment 14 was performed with an unlimited growth of recall. All three curves tended to approach an average score of 62 percent.

An even more difficult environment to predict was the statistically nonstationary sequence of symbols generated by classifying each of the increasing positive integers as being prime (represented by the symbol 1) or nonprime (represented by the symbol 0). That is, the environment consisted of the sequence 01101010001 ..., where each symbol depicted the primeness of the numbers 1, 2, 3, 4, 5, 6, 7, 8, 9, 10, 11, ..., respectively. Although in principle this deterministic environment should be perfectly predictable, no finite procedure is known for accomplishing such prediction (see Hardy and Wright, 1962, p. 6).

Figure 1.16 indicates the prediction score achieved against this nonstationary environment (experiment 15). After the initial transient, the prediction score gradually increased to reach 78 percent correct at the 115th symbol of experience, then remained essentially constant for the next 100 symbols. Examining the best-evolved finite state machines over successive predictions proved interesting. The 200th prediction was accomplished with a four-state machine, and the 201st with a three-state machine having only two output symbols of 1. The next two predictions were made by three-state machines, each of which had only a single output symbol of 1. All subsequent predicitons were made by a one-state machine with both output symbols being 0, which was due to a major "change of policy."[4] As a result, the prediction score once again gradually rose to reach its final score of 81.9 percent after the first 719 symbols of experience. This change of policy reflected the fact that prime numbers become less frequent as the experience grows. The results were obtained with a penalty for complexity of 0.01 per state, a population size of five machines at each iteration, and 10 rounds of mutation and selection between each next prediction of a new symbol. Figure 1.17 shows the policy in terms of the change in the number of states of the predictor machines.

To make the problem more interesting, we increased the length of recall and changed the goal so as to have greater payoff for finding a rare event. Under the new

---

[4]It was surprising, and perhaps significant, to note that prior to the change of policy, all of the missed primes were second members of twin primes.

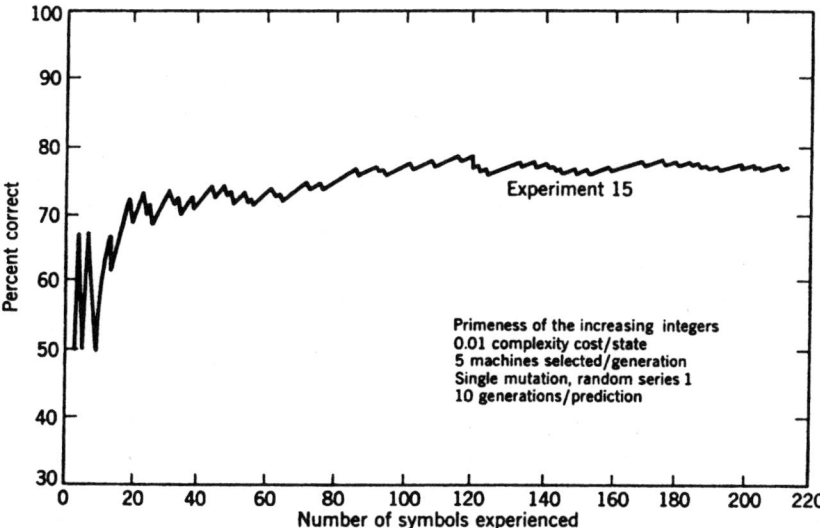

Figure 1.16 The cumulative percent correct for the evolutionary program to predict the primeness of the increasing integers. After the 201st prediction, the best predictor was a one-state machine that always output "nonprime" (from Fogel et al., 1966, p. 37).

rules, the credit received for correctly predicting a 1 was equal to one plus the number of 0's which preceded that 1, and similarly the credit received for properly predicting a 0 was equal to one plus the number of 1's that preceded that 0. This criterion, which places greater value on the prediction of rare events, reduces the prediction problem to the more usual one for an environment where 0's and 1's are equally likely.

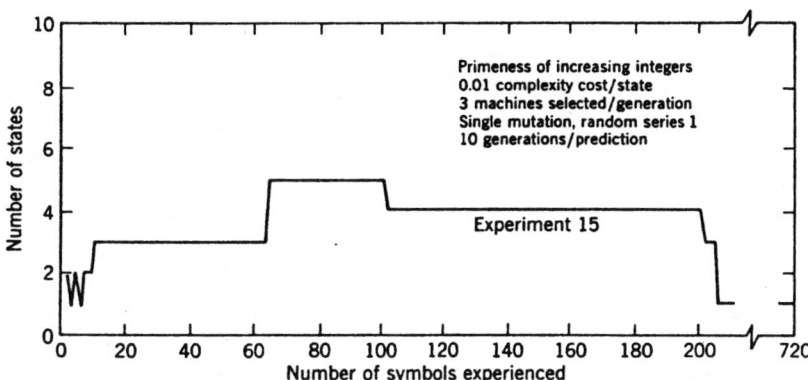

Figure 1.17 The size of the evolving finite state machines when predicting the primeness of the increasing integers. At the 201st prediction, and beyond, the best predictor became a single-state machine (from Fogel et al., 1966, p. 38).

In the first 150 symbols of this experiment 16, there were 30 correct predictions of primes, 37 incorrect predictions (false alarms), and 5 missed primes. In the sequence of experience from 150 to 547, there were 65 correct predictions and 67 incorrect predictions (false alarms). In other words, only 5 of the first 35 primes were missed—and none of the next 65 primes was missed. Analysis of the results indicated that the algorithm rapidly "learned" to recognize numbers that are divisible by 2 or 3 as not being prime. Some recognition that numbers divisible by 5 are not prime was also indicated. Here, again, it is worth restating that *the evolutionary algorithm nowhere contained any information concerning the meaning of "prime," nor any instruction to search for 1's in preference to 0's.*

It is not uncommon to encounter problems of predicting the occurrence of rare events in the real world. The search for "bonanza" and the concern for "catastrophe" constitute an essential part of everyday life. In many situations there is an uneven balance of cost for different kinds of failure. An example from the mid-1960s still holds today: The safety of future astronauts may well depend on an ability to forecast sudden bursts of intense solar radiation. Certainly false alarms are costly in that they require spacecraft crew to leave their normal assignment and remain within a sheltered area. However, the cost of a missed burst of radiation is far greater; it may constitute disaster for the crew and the mission.

The considerable success achieved in predicting whether or not the next integer is a prime number does not testify against the essential difficulty of the traditional mathematical problem of finding a single formula that would include only the prime numbers. The evolutionary programming approach did not search for such a single formula. In fact its primary advantage rests upon its allowing a different logical pattern for prediction as each new symbol is considered. Thus the evolutionary procedure provides a sequence of predictive logics that comprise the best models found thus far within the available expenditure and search capability. In the case of the prime number environment, it should be expected that the sequence of successive best models would tend to differ at each iteration in view of the nonstationarity of the generating function. The only necessary consistency among the sequentially best models lies in the fact that they were derived by the same evolutionary process.

A matter of interest in contemporary studies is the evolutionary program's ability to perform sequence prediction as compared with human subjects. Flood (1962) reported on seven human subjects who were required to make 450 consecutive binary choices (decisions). After each choice the subject was informed as to the success or failure of his last choice, earning one cent for each success. Each subject repeated the experiment once, but against a different "environment." The evolutionary program was exercised on the same sequences over 200 choices and in the 14 trials (two for each of the seven subjects) it outperformed the human half of the time. In a separate experiment Wolin (1963) reported on trials with college students who were asked to predict symbols in a sequence of 1000 binary symbols. Each subject was given one sequence and provided with a visual display of the last eight symbols and predictions at each point in time. Six environments were offered such as where odd symbols were equally likely to be *A* or *B* while all even symbols were *A*. Another varied the probability of an *A* or *B* on the odd symbol. More complex sequences involved statistical sequences of word elements (groups of symbols). Comparisons to the evolutionary program showed that when the finite state machines were provided with the entire observed sequence for recall, the evolutionary process consistently outperformed the human subjects but did not do as well when the recall was limited to the last 10 symbols (Fogel et al., 1966, p. 42).

Similar prediction experiments were conducted using eight-symbol alphabets (0, ..., 7) where the environment was subjected to noise. Probabilities for altering symbols by ±1 or ±2 points were incorporated (with addition to the symbol 7 or subtraction to the symbol 0 leaving these cases undisturbed), as well as "wild noise" which could change a symbol to any other symbol uniformly at random. The population was set at four parents with a maximum number of states of at least 15. The initial recall was 40 symbols, and a penalty for complexity of 0.01 per state was imposed in evaluating each finite state machine. Although repeating the experiments was time-consuming, the introduction of randomness always introduces questions of repeatability; therefore simple experiments involving only ±1 noise were repeated 10 times with the results depicted in Figure 1.18. As might be expected, the variability was seen to be an inverse function of the score attained: The higher the predictive ability, the lower was the variability.

Additional experiments were conducted predicting zero- and first-order Markov sequences in four symbols, as well as the more difficult case of predicting the sequences created by rank ordering the powers of 2 and 3 reduced modulo 8 or 7 (Fogel et al., 1966, pp. 48–53). Consideration was also given to predicting sequences with inputs that required a pair of symbols coupled with a single output symbol (see Fig. 1.19).

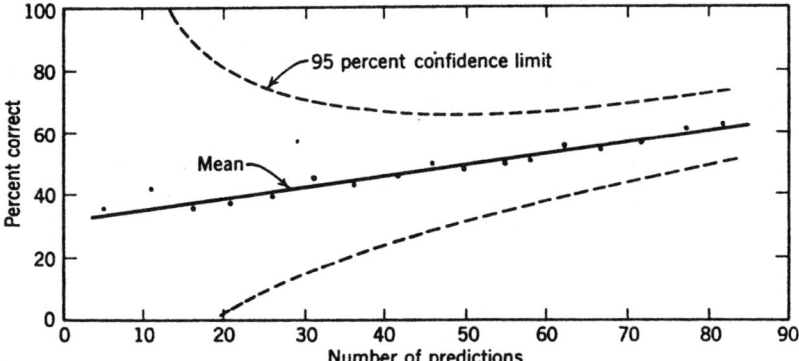

Figure 1.18 The mean and variability of 10 trials of the evolutionary program predicting the eight-symbol sequence (13576420)* corrupted by ±1 noise. As performance improved, the variability of the results decreased (from Fogel et al., 1966, p. 38).

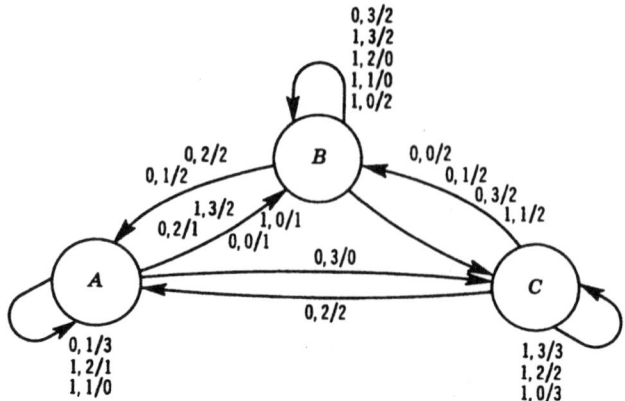

Figure 1.19 A three-state machine using symbol pairs for inputs and a single output (from Fogel et al., 1966, p. 54). A similar form of a machine was later used in evolutionary programming experiments with the iterated prisoner's dilemma (Fogel, 1995, ch. 5).

1.3 PATTERN RECOGNITION AND CLASSIFICATION

Situations often arise in which a sequence of measurements is examined in order to determine whether a pattern or some regularity exists within the data. An analogous situation is found in the first step of cryptanalysis: Is a message contained in this sequence of symbols? The problem of *detection* can be translated into the problem of finding sudden changes in the predictability of the data. A consistent inability to predict each next symbol leads to the suspicion that it is purely random. In contrast, a steady and significant prediction score may reveal the presence of an unchanging signal. Variability in the prediction score reveals that the data may contain a message.

Demonstrating an ability to predict the observed data immediately leads to the question: What is the nature of the signal? The answer may be found by way of the evolved predictor-machines. Each machine provides, in a sense, a description of either the periodic or the stochastic properties of the environment. Every finite state predictor-machine has a characteristic cycle that can be found by driving the machine by the sequence of its own output symbols. At some point the sequence of output symbols will become an endless repetition of the characteristic cycle. If the environment contains periodic properties, these should be found in the characteristic cycle.

Similarly each finite state machine may be viewed as a deterministic procedure that best typifies the statistical properties of the data in light of some payoff function. If the environment is considered to be of first order, with equal payoffs for each possible correct prediction and equal penalties for each possible error, then each state transition should correspond with the probabilistic transition that occurs with the highest conditional dependency. By the same token, the zeroth-order properties of the environment should be evident in terms of the relative frequency of the output symbols of the evolved predictor-machines. It is more difficult to estimate the second- and higher-order conditional probabilities of the environment directly from evolved finite state machines. In a sense, this information is stored in the machine because it has affected the sequence of selected machines; however, direct read out of this information is not always possible.

At one extreme, a repetitive environment may be viewed as the limiting case of a stochastic process in which the order equals the length of the period. Under such conditions the evolved machine will be cyclic, and the "conditional" dependency will be clearly evident in the state transition diagram. At the other extreme, where the conditional probabilities are nearly equal, it becomes difficult (if not impossible) to estimate these probabilities from the configuration of the evolved machine. Of course, if such information appears to have a particular value, the evolutionary program could be written to consider extended finite state machines, machines that designate each transition in terms of the last few symbols whose number defines the highest-order conditional properties to be expressed.

But pattern recognition is only a first step. All too often the search for artificial intelligence is reduced to a simple attempt to build automatic pattern recognition devices. This is just one of the capabilities that identify the existence of intellect. First, there is the problem of detection: Is there a signal in the noise? Then there is the problem of discrimination: If there is a signal, what is it? With this accomplished, it now becomes reasonable to inquire: Is this signal one of those already known? So literally the problem is one of re-cognition. In practice, the problem is given larger scope: To which of the known signals is this signal most similar? This becomes the problem of pattern classification.

We can now proceed to the question: Can an evolutionary program recognize and classify patterns in a manner similar to that of an average human operator? To make the question specific, 16 different amplitudes of time patterns were generated (reported in Fogel et al., 1966). They were naturally classified into four groups based on their waveform similarity. Each pattern was generated in 35 possible amplitudes at 125 equally separated points in time by recording the rectified envelope of a waveform that resulted from adding Gaussian noise to the four different primary events (see Fig. 1.20).

These amplitude data were encoded in an eight-symbol alphabet for use in the available computer program in order to equalize the marginal probabilities of each symbol. For simplicity, each pattern was started with the first excursion from the symbol 0 and terminated after 45 symbols had appeared. The encoded data are shown in Figure 1.21. Note that some subjective similarity remains among members of each successive quadruplet.

The previously described eight-symbol evolutionary program was used to predict each next symbol in an unending repetition of each of these patterns. There was (1) no penalty for complexity, (2) 10 generations prior to each prediction, and (3) a "magnitude of the difference" error cost matrix specification of the goal. Table 1.2 indicates the average error in "predicting the past" for each of the predictor-ma-

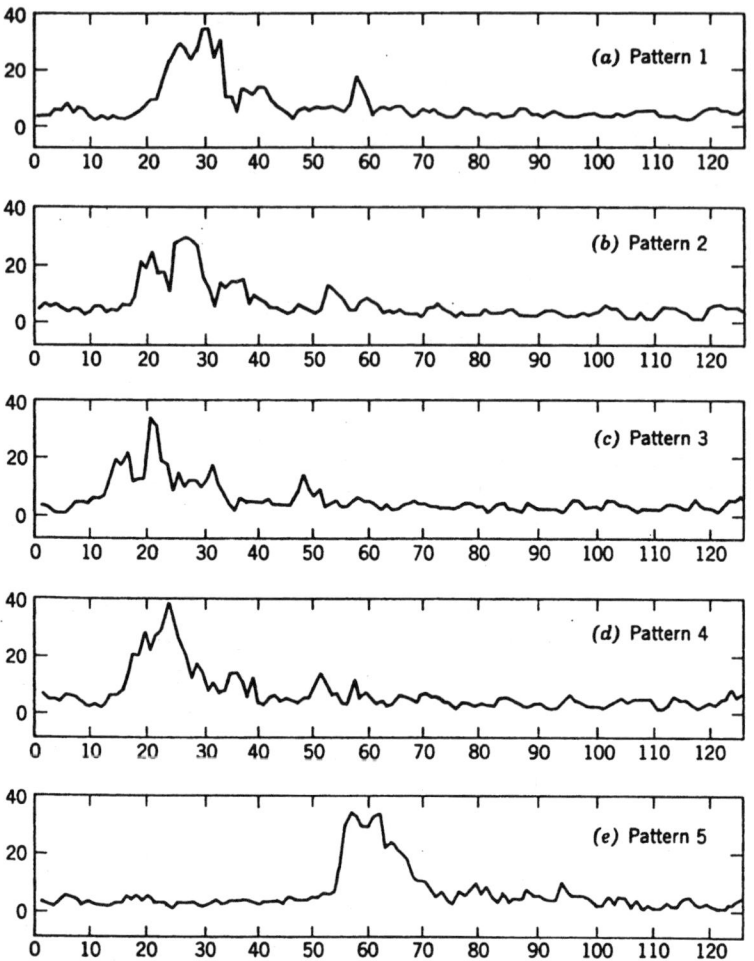

Figure 1.20 Sixteen patterns (*a–p*) generated in 35 possible amplitudes at each of 125 points in time. Each pattern is the result of a filter applied to one of four primary events corrupted by Gaussian noise (from Fogel et al., 1966, pp. 76–78).

chines after the first 50, 100, 200, and 400 predictions. The greatest portion of the learning was seen to take place early in the evolution.

Each evolved predictor-machine was a characterization of the pattern against which it was evolved. Similarity of the patterns might be deduced by noting similarities in the machines. However, this is not easy. The complexity of the machines makes this difficult even if a suitable set of parameters for making such judgments were available. A more natural way to accomplish the comparison is to allow each evolved predictor-machine to predict the remaining patterns. The similarity between certain patterns should become evident by the nearness of prediction scores. Each

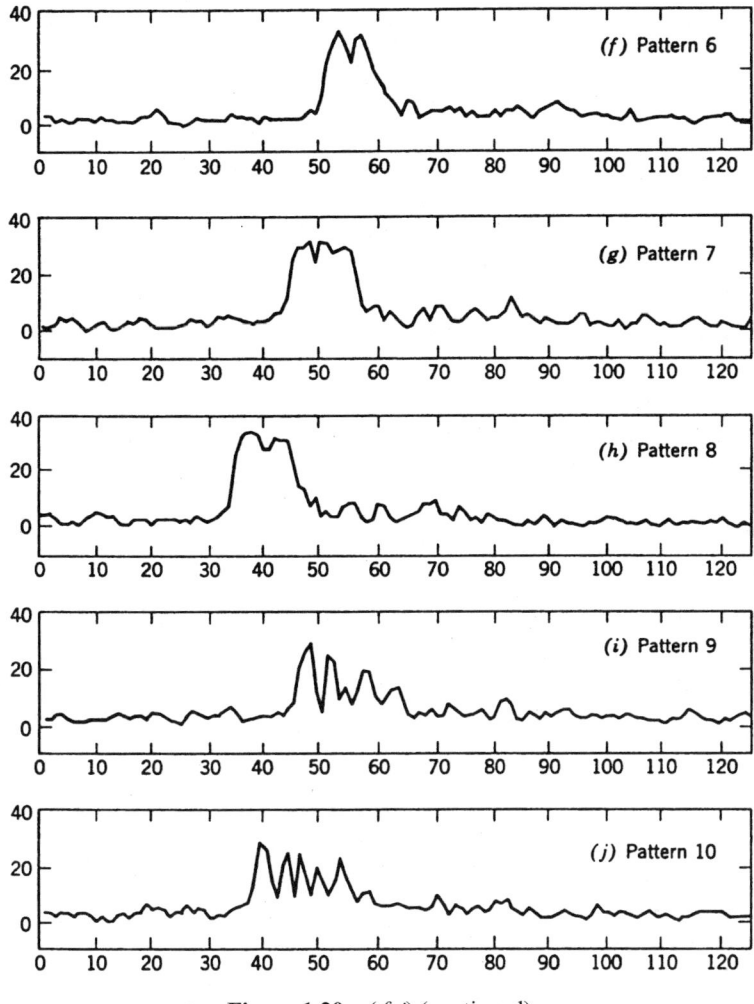

Figure 1.20 (*f–j*) (continued).

predictor should score best on the pattern against which it was generated and nearly as well on "similar" patterns, but poorly on all other patterns.

In the next set of experiments, one of the predictor-machines evolved against patterns 1, 2, and 3 (i.e., three machines were selected, one from each case) was used to predict each of the 16 patterns using a squared error cost matrix. Table 1.3 indicates the results, with the average error of prediction tabulated over 45 and 90 symbols. As expected, there was little difference between these sets of data. Each predictor-machine easily recognized its own pattern, but the remaining scores showed that none could classify the patterns in the desired manner.

Note that the goal of comparison expressed an equal value for each correct prediction and an equal cost for errors of the same magnitude. On inspection it is evi-

24 GENESIS

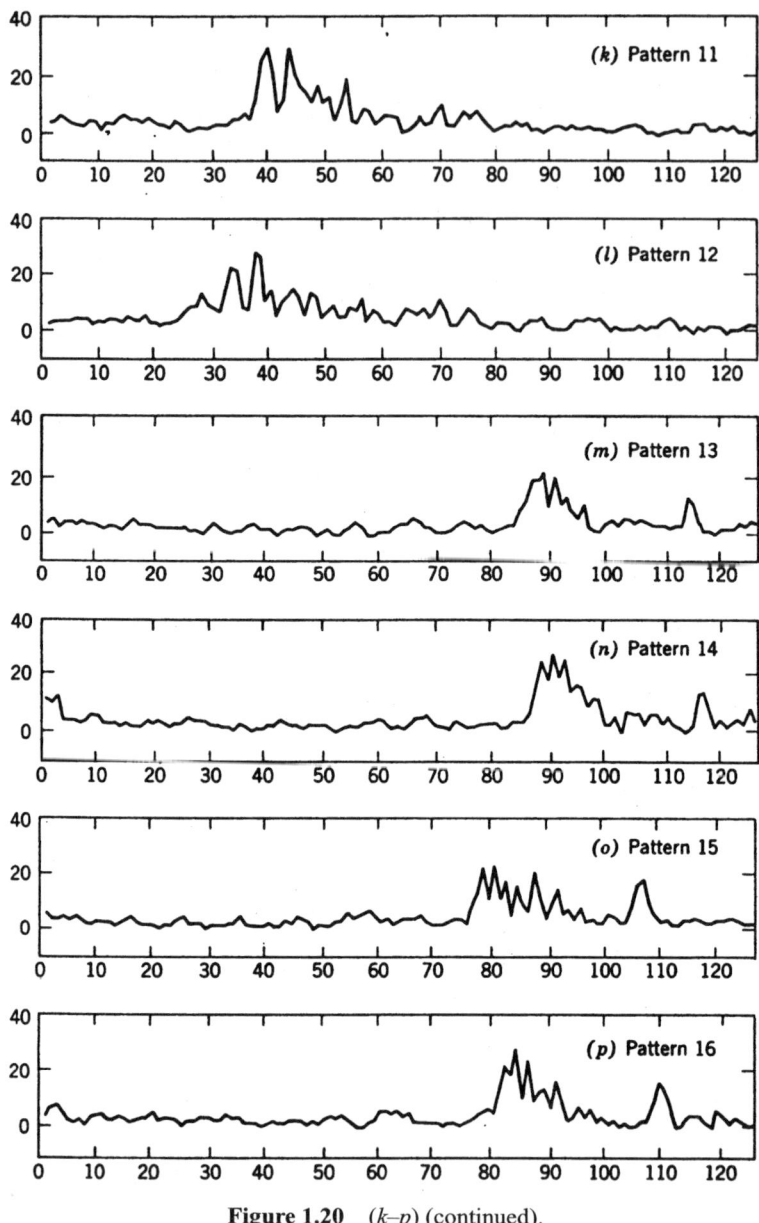

Figure 1.20 (*k*–*p*) (continued).

dent that the predictor-machine judges similarity in a different way from how a human might approach the problem. For instance, an average observer is apt to characterize a pattern by the height and position of its peaks. If pressed further, he might consider the number of peaks and valleys, their relative magnitude, position, and so forth. But there is no demand that the evolutionary program emulate human behav-

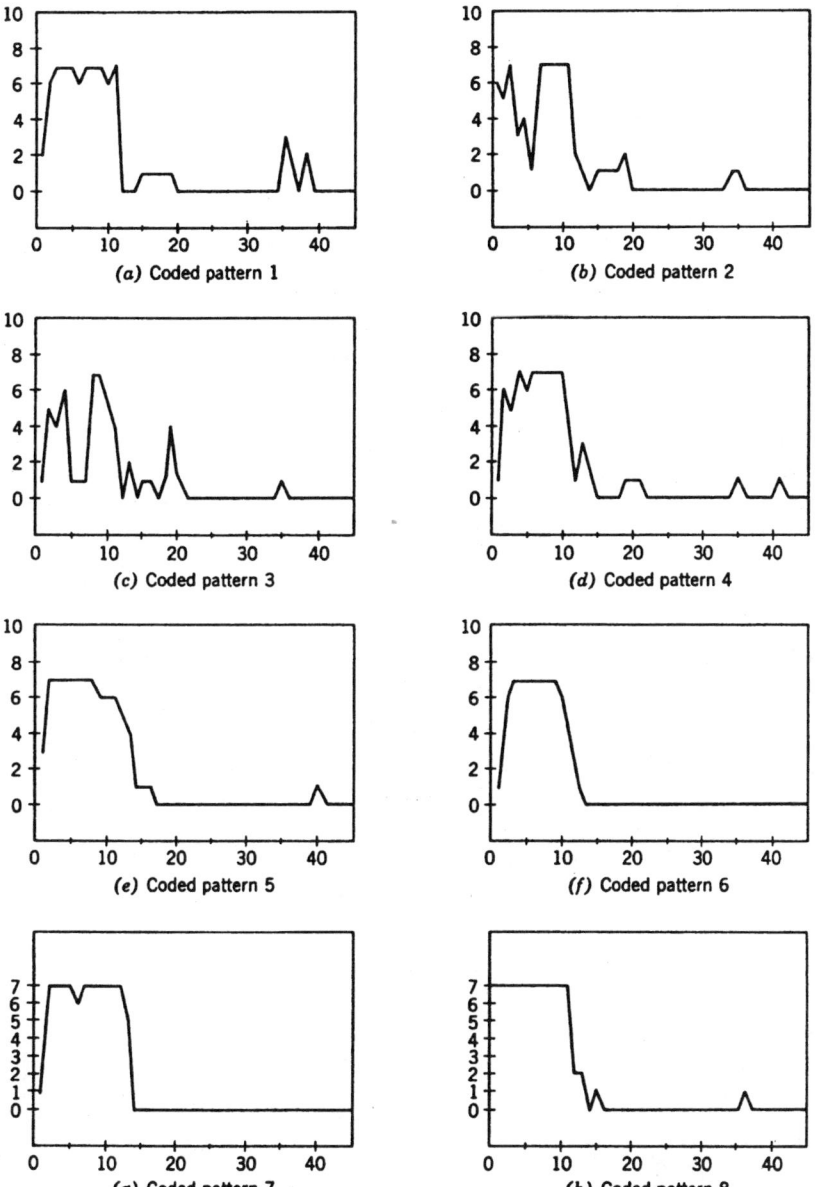

Figure 1.21 The patterns from Figure 1.20 (*a–p*) encoded in eight symbols (from Fogel et al., 1966, pp. 80–81).

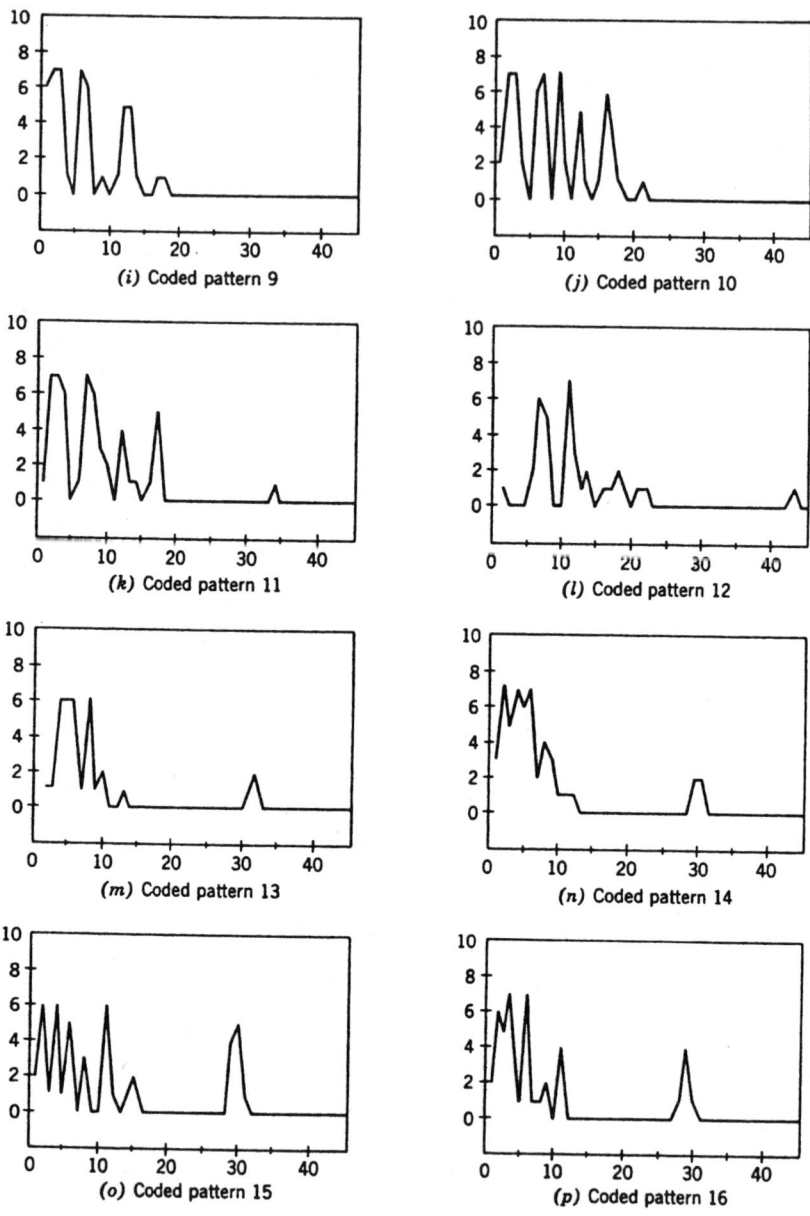

Figure 1.21 (*i*–*p*) (continued).

Table 1.2 Results of applying the eight-symbol evolutionary program to predict each next symbol in an unending repetition of the patterns in Figure 1.21

Pattern	Recall Length			
	50	100	200	400
1	0.330	0.172	0.158	0.165
2	0.307	0.306	0.211	0.145
3	0.228	0.195	0.091	0.056
4	0.248	0.209	0.111	0.114
5	0.106	0.091	0.057	0.045
6	0.043	0.051	0.051	0.047
7	0.082	0.041	0.048	0.050
8	0.268	0.267	0.288	0.276
9	0.185	0.163	0.172	0.149
10	0.207	0.141	0.088	0.070
11	0.184	0.082	0.078	0.052
12	0.312	0.107	0.090	0.072
13	0.122	0.102	0.091	0.081
14	0.103	0.126	0.110	0.109
15	0.165	0.176	0.152	0.136
16	0.063	0.068	0.067	0.065

Note: The table indicates the mean error for each of the predictor-machines after the first 50, 100, 200, and 400 predictions.

Table 1.3 Results of using the FSMs trained on the first three patterns to predict the sequence of symbols from all 16 patterns

	Predictor-Machine 1 Evaluated over		Predictor-Machine 2 Evaluated over		Predictor-Machine 3 Evaluated over	
Pattern	45 Symbols	90 Symbols	45 Symbols	90 Symbols	45 Symbols	90 Symbols
1	0.250	0.292	4.727	5.865	12.136	12.045
2	4.727	5.315	0.023	0.427	5.159	7.787
3	4.841	4.697	4.864	4.820	0.023	0.034
4	4.500	3.629	5.659	6.00	11.159	11.258
5	2.796	3.348	5.886	5.112	5.432	8.843
6	1.091	1.539	3.136	3.933	9.068	9.551
7	1.296	1.697	3.364	3.112	11.886	11.933
8	1.432	2.753	2.318	2.292	12.091	12.506
9	3.773	4.685	5.000	5.348	4.682	5.315
10	6.273	6.787	8.546	7.225	6.068	6.472
11	5.250	5.607	5.136	4.798	3.818	4.292
12	3.000	2.978	3.250	3.225	4.727	4.573
13	4.432	4.258	3.136	3.337	3.796	3.764
14	5.341	5.382	3.818	4.112	6.727	6.753
15	4.455	4.854	4.386	4.214	5.046	4.989
16	3.682	4.090	2.636	3.416	5.682	5.663

Note: In each case the FSM "recognizes" (i.e., has a much lower error score on) the sequence generated from the pattern that it was trained on as compared with the other patterns.

ior in performing the same task. In fact it is this very constraint that has limited the advancement of artificial intelligence for over 30 years.

1.4 CONTROL SYSTEM DESIGN

Attention was previously given to such problems of data reduction as *detection* (is there a signal?), *discrimination* (if so, what is the signal?), *recognition* (has this signal been seen before?), and *classification* (if not, which of a set of signals is it most like?). Only at this point is it appropriate to address a more significant problem: the problem of control. (A situation has been characterized, now what should be done about it?) In essence, the prior problems were of interest only in that they might precede steps toward a solution of the problem of control.

A given system is referred to as the *plant*. Situations often arise in which it is desired to make the plant perform in a particular manner. If the desired behavior is not expected to result from the natural interaction of the plant with its environment, the problem is then to manipulate the available input parameters of the plant in a way that will bring about this performance, despite the essential dynamics of the plant and any internal or external disturbances that may occur. The problem is more difficult if knowledge of the nature of the plant and the disturbance is incomplete. The extremely difficult case is where the desired performance can only be measured when it is nearly complete, or where the evaluative criteria have been only partially specified.

This very general view of control system design problems is offered only to emphasize the limitations of classic control theory. Conventional control techniques can be used if it is desired to make a linear plant satisfy certain constraints on its steady-state or transient response. In the mid-1960s little attention was devoted to the design of control systems for distributed plants (plants that require an infinite number of descriptors to fully define their state). Certain effects of stochastic disturbance were examined in great detail, but the effective control design procedure reduced such problems to their deterministic "equivalent" by focusing on average disturbance.

Such analytic design procedures become inadequate when there is incomplete knowledge of the plant to be controlled. Iterative techniques may be called upon, but their success depends on the availability of suitable evaluation criteria that can be referred to at each iteration. The problem becomes more formidable when the performance of the plant can only be evaluated at particular points in time. In the limit, where the time between samples is small, the problem becomes unsolvable because the time remaining to achieve control runs out.

In this discussion, attention will be restricted to plants that operate only at discrete points in time and have input parameters and performance that can be measured within a finite set of symbols. The parameters are taken collectively to comprise the input variable. The performance of the plant is described in terms of a sequence of symbols that constitute the controlled variable. The plant is presumed to be in one of some finite number of discriminably different internal states during each transduction. The worth of each symbol of the controlled variables is measured in terms of a completely specified payoff matrix that relates actual performance to desired performance at each point in time. The elements of this matrix may be time dependent or may be functions of the worth of previous performance. The desired overall performance may be that which will maximize the worth over some sequence of symbols received from the environment.

Suppose the intent is to have a given, but unknown, plant optimally track an arbitrary reference signal in the sense of satisfying any well-defined goal. To accomplish this end, an evolutionary control system operates in the following manner: The sequence of symbols that has most recently driven the given plant, M, is used to drive an arbitrary initial finite state machine M'_1, which is viewed essentially as an estimate of the plant M. Each output symbol of M'_1 is compared with the corresponding symbol that has emerged from M. The penalty for any discrepancy is then found by referring to the error cost matrix that is an explicit statement of the purpose of the control system. The cumulative penalty over the sequence of symbols that comprises the recall provides a measure of the inadequacy of M'_1 as a representation of M in the light of this goal and the experienced environment.

An offspring of M'_1 is now produced through mutation. This new offspring M'_1 is also exposed to the same sequence of input symbols, and its worth as a representation of the plant is similarly scored. If this discrepancy score is less than or equal to that of the parent, this offspring survives to become the new parent. If not, it is discarded, and a new offspring is generated. In this manner nonregressive evolution proceeds to identify a succession of finite state machines that are better representations of the plant M. The method may be easily extended to include a population of more than one parent candidate representation.

At any point in time, the most recently evolved best machine can be examined in order to determine that particular symbol that would serve as the most appropriate input for the plant, that is, given the present state of the machine, which transition will yield an output symbol contributing the least penalty in terms of the error cost matrix. After the most appropriate input has been identified, that input symbol is used to drive the plant. If multiple transitions in the evolved machine are found to be equally suitable, then an arbitrary rule is applied to result in a selection of a particular input (i.e., this could be to choose one particular satisfying input at random). If additional sophistication appears justified, that transition is selected which leads to a state that offers the possibility of a wider selection of output symbols or of the particular output symbol that will next be desired, if that symbol is known.

Fogel et al. (1966, pp. 86–98) described experiments with such a protocol in order to control a one-state, a three-state, and a five-state machine. The number of experimental trials that could be conducted was quite limited, however, due the computational requirements and the available computing hardware. To better illustrate the concept, these original experiments have been recapitulated and presented in Fogel and Chellapilla (1998). One typical result is shown in Figure 1.22 where the evolutionary program identified and controlled a five-state machine with a different transfer function in each state.

30 GENESIS

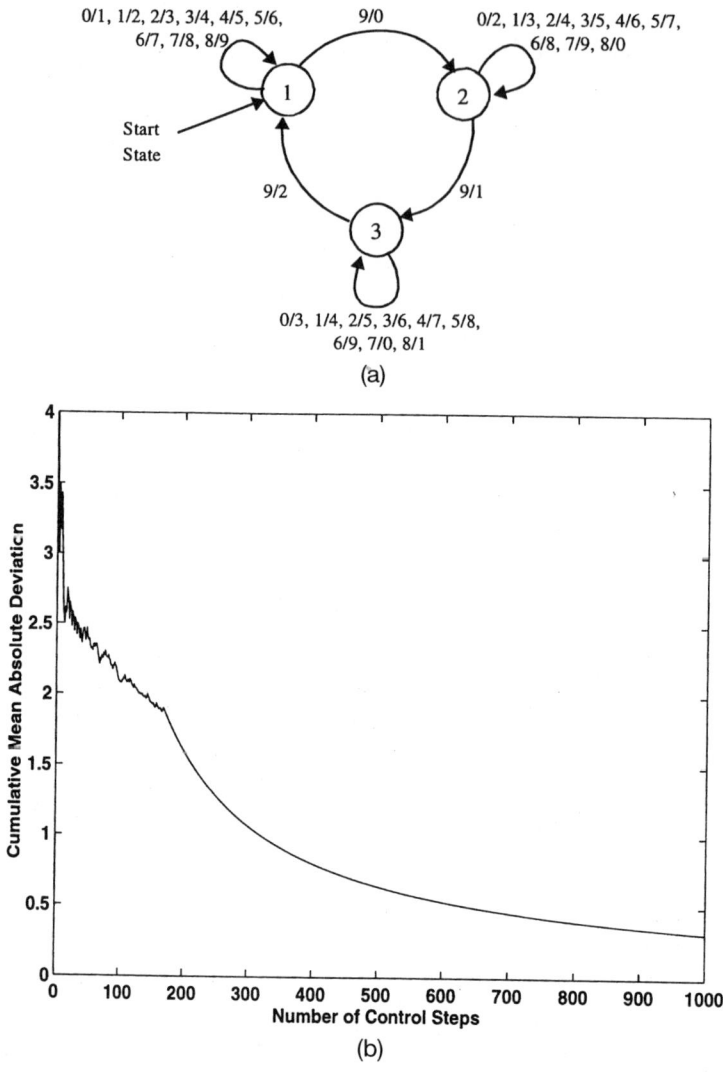

Figure 1.22 Fogel and Chellapilla (1998) recomprised the original evolutionary programming control experiments in Fogel et al. (1966) and applied the technique to controlling n-state FSMs with 10 symbols of the form shown in (*a*), where the output given an input symbol was the state number added to the input symbol. If the output would have exceeded 9, it was wrapped back around 0 (e.g, in State 2, the input 9 yields an output of 1). All transitions lead back to the same state except on the input of the symbol 9, where the machine transitions to the next state incrementally. The goal was to identify the machine using the evolutionary program and control it to output the symbol 5. Note that any such machine is completely controllable with respect to this task. (*b*) Typical control error of the evolutionary program when controlling a five-state FSM under the above conditions. The cumulative mean absolute error decreases as the evolutionary program learns the dynamics of the plant. After the 168th control step, the evolutionary program discovers a logic that can completely control the plant without error. For further details, see Fogel and Chellapilla (1998).

1.5 EXTENSIONS OF EARLY EVOLUTIONARY PROGRAMMING CONCEPTS

The early experiments with evolutionary programming were constrained by the available computer speed and memory. Several ideas were considered as potentially important but were not given sufficient experimental attention. These included (Fogel et al., 1966, pp. 21–25):

1. A suitable choice of the mutation noise may increase the efficiency of evolution. For example, an increase in the probability of adding a state generates a wider selection of larger machines, which is beneficial when the environment is complex. The probability distribution over the modes of mutation can be made to depend on the evidence acquired within the evolutionary process itself. Thus an experienced greater relative frequency of success for, say, changing the initial state might be made to increase the probability of this mode of mutation.
2. Intuitively, the best surviving parent is most likely to produce superior offspring, and should receive the most attention in mutative reproduction. However, lower-ranked parents may be regarded as insurance against gross nonstationarity of the environment. The distribution of mutative effort may well be in proportion to the normalized evaluation scores.[5] All of the retained machines need not lie on the slopes of the peak identified by the best machine. Saving a collection of best offspring may maintain a "cognizance" over several peaks, with the relative search effort being distributed in proportion to the expectation of significant new discoveries.
3. The recombination of individuals of opposite sex appears to benefit naturally evolving organisms. By analogy, worthwhile traits that have survived separate evaluation might be retained by combining the best surviving machines through some genetic rule, mutating the product to yield offspring. This mating does not need to be restricted to the best two surviving "individuals." The most obvious genetic rule, majority logic, only becomes meaningful with the combination of more than two machines. Each state of a majority logic machine is the composite of a state from each of the original machines. Thus the majority machine may have a number of states as great as the product of the number of states in the original machines. Each transition of the majority machine is described by the input symbol that caused the respective transition in the original machines, and by the output symbol that results from the majority element logic being applied to the output symbols from each of the original machines. To illustrate, consider the three machines shown in Figure 1.23. The majority logic machine in Figure 1.24 can be created by examining each triple of possible states and the majority response to each input stimulus.

---

[5]This idea was principally the same as the mechanism of roulette wheel selection that was later popularized in genetic algorithms.

32 GENESIS

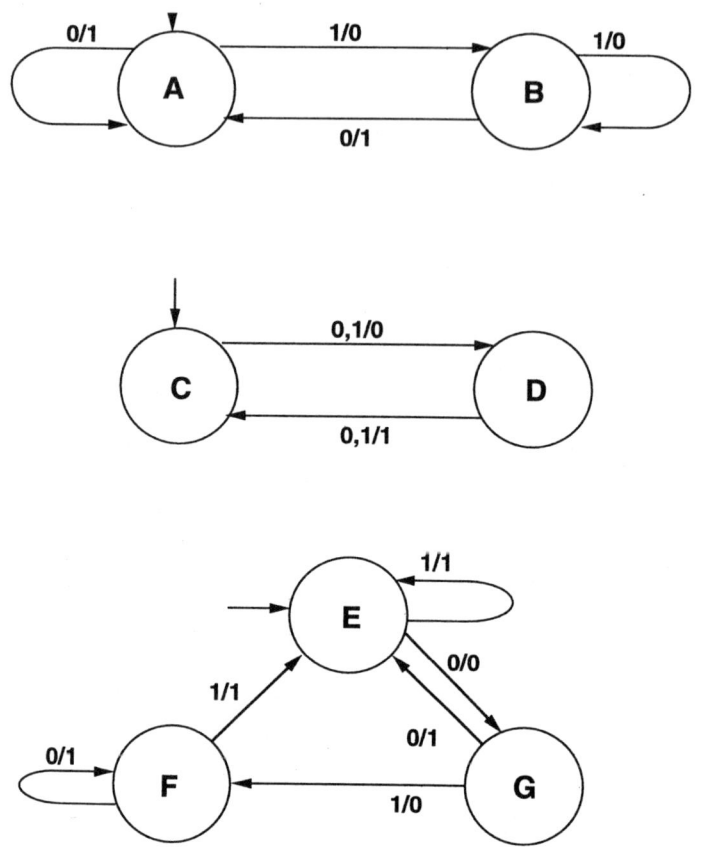

Figure 1.23 Three machines for majority logic (from Fogel et al., 1966, p. 22).

Each of these ideas was worthy of consideration but was too computationally burdensome for sufficient investigation in the 1960s. For example, the size of a majority logic machine grows with the product of the number of states of the individual parents, thereby quickly exceeding the available computer memory. Adapting mutation rates was later considered by students of Don Dearholt at New Mexico State University in the 1970s. Proportional selection appeared to be a luxury that might slow down convergence, so efforts in probabilistic selection in evolutionary programming were dormant for two decades before being studied again in Fogel (1988).

1.6 COMPETITIVE GOAL-SEEKING

To some extent, all decision-making entities are themselves affected by the response of the real world. That is to say, in principle, it is impossible to react to the observed

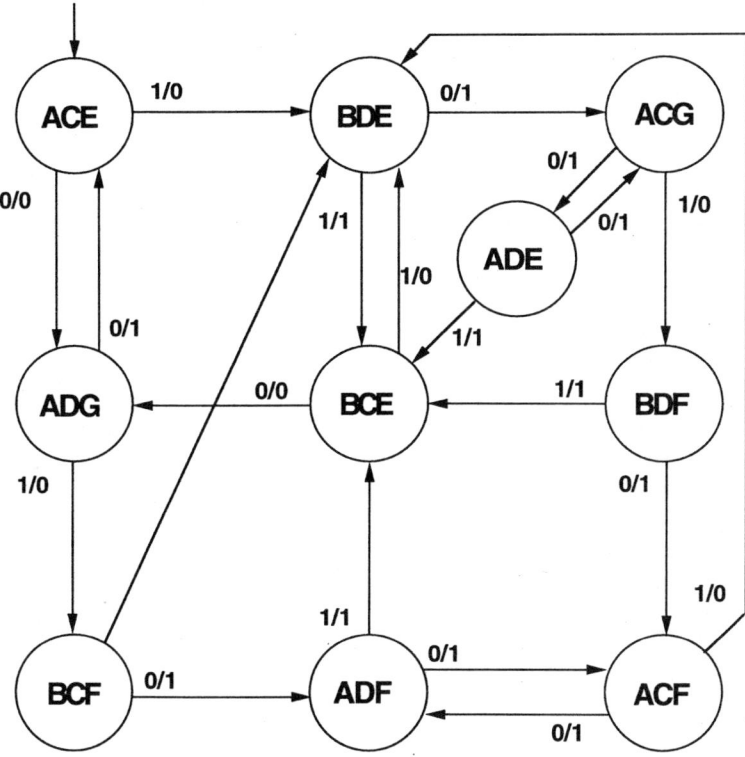

Figure 1.24 The resulting majority logic machine (corrected from Fogel et al., 1966, p. 23).

environment without affecting it, at least to some degree. In the previous discussion it was tacitly assumed that the influence of the decision maker on the environment is negligible. In other words, the communication link between the environment and the observer was considered to be essentially unilateral. At this point it is of interest to consider the more general situation of bilateral communication wherein the decision-making entity both affects and is affected by its environment.

In his classic work John von Neumann considered the problem of devising optimal strategies in competitive game playing. He provided a solution for the player who desires to gain maximum payoff at minimum risk, finds himself in a zero-sum situation, and anticipates an unending game. Although considerable effort has been devoted to expanding the domain of applicability of game theory, it remains true that most real-world competitive situations present problems that fall well outside the very special class that is presently amenable to solution. Evolutionary programming offered a novel and different means for devising strategy over an extremely wide domain of competitive solutions.

As a natural extension of the previous discussion, consider the situation wherein an organism exists with an external environment. This organism is equipped with sensory and motor capability, a memory, and an ability to process the sequence of symbols that are sensed in order to determine some desired response in the light of the given goal. The organism may be limited in terms of its ability to translate the result of such decisions into a corresponding action. That is to say, the physical properties of the motor capability may be viewed as forming a "plant" that serves to translate the results of a decision into a controlled variable that is the organism's actual response to its environment. The worth of each response is indirectly measured by the organism in terms of the payoff associated with the future symbols it will sense. The measure of its goal-seeking ability is in terms of the cumulative payoff over the sequence of the sensed symbols experienced thus far.

The fundamental problem of the organism is identical with that of system control where it is desired to have an unknown and possibly limited plant track a reference signal, the cost of errors being expressed in terms of some given goal. As indicated earlier, this problem can be addressed through evolutionary programming.

A more difficult situation occurs if the plant can change its logical structure. Under such circumstances it is important that there be a sufficient sample of input and output symbols in order to permit the construction of an adequate model of the unknown logic of the plant. It is difficult to estimate the size of such a sample, since it largely depends on the number of symbols in the alphabet and on the degree of change in the logic. And yet, in principle, the evolutionary approach remains valid in that it affords a means for utilizing the available information to the limit of the computational capability.

To generate the proper output symbol that is intended to influence the external plant, the organism must generate a model of its own plant (self-reference). Greater difficulties arise if the sensory capability of the organism is so noisy that a random disturbance is imposed on the sensing of its own response. Such noise may be expected to degrade the capability of the organism to effectively model itself, and thus to optimize its own goal-seeking performance. Here again, though evolutionary programming appears to be a valid approach, the required sample size also increases as the uncertainly about the organism's own actions increases.

Even greater difficulty is introduced if the organism cannot directly sense its own response, that is, if the environment intercedes between the controlled variable and the sensed variable (e.g., late in his life, Beethoven could not hear the music he created—his judgment was limited to mentally simulated sounds). Hopefully, the environment offers some relatively simple logic connecting these controlled and sensed variables, a logic that changes little with time. Placing the organism in a noisy environment will handicap its ability to erect a model of the composite of its own plant and the logic of the environment in order to proceed further with its goal-seeking as measured on the sensed variable. Unfortunately, real-world situations usually include just such difficulties. The environment might contain other organisms that in turn respond to the activity of the subject organism. If the sensory capability of the subject organism is severely restricted, it may view the composite of such response in terms of a single descriptor at each point in time. Its task is then to form an adequate model of the entire environment as seen in toto. If, however, the organism has sufficient sensory resolution, it may proceed to model each of the various "aspects" of its environment separately, and in this manner it may choose a more suitable course in the light of its goal (see Fogel, 1963, ch. 11).

It might be that these other organisms are also goal-seeking entities and modify their behavior as a function of the degree of success of the subject organism. Their performance then might enhance or degrade the situation of the subject organism, but in any case, its competition exists only on an absolute scale (i.e., it lives to reproduce or dies). Such an organism is engaged in competition only at the lowest level; it cannot make use of devices such as bluffing, deception, collusion, and others that might serve to enhance its position, both on an absolute and on a relative basis. If its own behavior happens to benefit some other organism, there exists natural coalition, but such opportune activity is inadvertent.

For the organism to rise to a higher level of behavior, it would have to possess an internal ability to separately sense its own stimuli and response in order to construct models of itself (these being distinct from the models of itself that may be constructed as reflections of its own behavior as seen in terms of its response of the environment—we know ourselves in large measure in terms of other people's responsive behavior). That is to say, a portion of the organism would have to be set aside for the purpose of representing the "self" in terms of some given goal (which may be different from the goal of the external performance of the organism). Note that the separation of such an information-processing unit denies the organism the ability to model itself completely, in the sense that "the cat can never taste its own tongue." It is of course possible to set aside a section of this modeling capability for modeling the remainder; however, such successive sectioning soon finds an end in its limited utility under inadequate resolution.

In order for a model of self to be useful at any level, it must be compared with a model derived as a reflection of the environment. The comparison of these different kinds of models provides an ability for the organism to "describe itself" to its external environment. This is the beginning of self-awareness, a basis for conscious behavior and communication at a higher level than "conditioned response."

In many situations it is wise to assume that the environment contains at least one intelligent adversary, an entity that will modify its transduction so as to inhibit, if not prohibit, its control. Again, evolutionary programming is an appropriate way to proceed, except that here care must be taken to avoid the use of deterministic "inner loops" (i.e., internal feedback procedures) which might ordinarily be adopted in order to increase the efficiency of evolution. Whenever fixed rules of behavior are imposed, there is always the danger that the plant may in turn model the "controller," at least to the extent that it can detect such regularities and, as a result, adopt an obverse policy. In general, greater efficiency in effecting control is gained at the cost of a loss in security.

Surprisingly enough, the additional "complication" of a bilateral communication channel offers a distinct advantage to the decision maker. With an ability to affect the behavior of the environment comes the possibility to "interrogate" the environment by selecting stimuli and analyzing the observed response. Hopefully, there may appear some highly certain stimulus-response relationship, so the decision-maker can choose to avoid those actions that are expected to reduce the achievement of its goal and take those other actions that have a significant likelihood of resulting in some benefit. In any case, the greater is the number of "known" stimulus-response pairs of the environment, the greater the probability of choosing sequences of action that will lead to a more favorable situation.

In general, then, the competitive decision-maker should have an ability to construct (i.e., *evolve*) representations for the logic that can be used to "explain" the observed response-behavior of the environment. On the basis of such a model, the environment can be made to behave more predictably and in a manner more favorable to the decision-making entity (for further discussion, see Fogel et al., 1966, pp. 102–107).

1.7 SOME IMPLICATIONS

1.7.1 Evolution and the Scientific Method

It is useful to consider some implications of the preceding philosophical and experimental progress. Primary of these implications is that the scientific method is an evolutionary process. It is an iterative process that facilitates the gaining of new knowledge about the nature of an observable environment. The evolutionary aspects of this process become apparent upon examining the process in some detail.

Initially the investigator chooses to estimate the value of some unknown aspect of his environment. Having established a goal, he collects whatever empirical evidence may be available from previous observations (removing, of course, those systematic errors known to exist). Although he recognizes the fact that these data are not without error, he presumes them to be true and proceeds to generate a model of the environment that is consistent with as much of these data as possible. This process of generalization is called *induction*. It constitutes the creative step of the scientific method. The models that result are given various names—hypotheses, conjectures, theories, representations, rules, language—depending on the context.

A model is useful only if it provides a basis for an estimate of the specific unknown data point under consideration. Finding that specific estimate from the model is a deductive specialization. To facilitate this process, it is often worthwhile first to perform some deductive transformation of the model, that is, manipulating it into some more tractable form called a "general solution." Of course, the scientific method remains incomplete until there has been an attempt to verify or deny the validity of this estimate of the unknown data point through an independent observation of the environment. These logical steps are indicated in Figure 1.25. In brief, the scientific method consists of the induction of an appropriate model, the inductive inference of a specific unknown data point, and an independent measurement intended to verify or deny this inference; these steps are iteratively applied to the expanding base of known data points.

With each measurement comes a benefit. Even if the estimate derived from the model is not verified, the effort is not in vain, for a new data point is now known. This enlarges the set of empirical evidence that may be assumed "true" for succes-

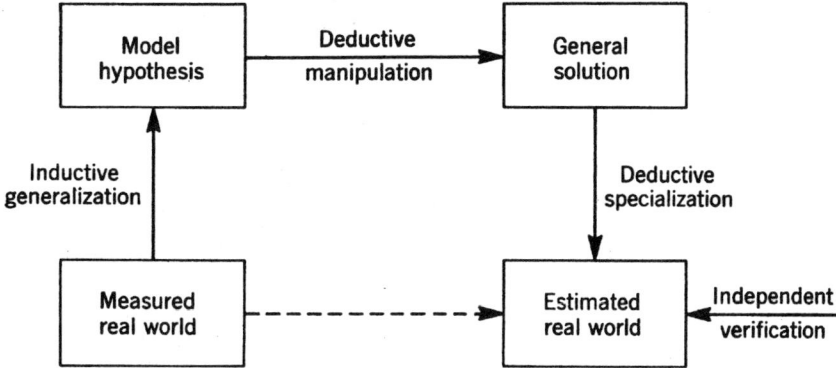

Figure 1.25 The scientific method (from Fogel et al., 1966, p. 109).

sive iterations. With an increased sample size, it should be possible for the investigator to create a better model, relative to some particular environment. As the knowledge of this domain grows, there is greater likelihood that each successive estimate will be verified.

An additional and deeper benefit accrues whenever an estimate of the unknown is verified. Then it becomes reasonable for the investigator to increase his belief in the value of that specific model as a valid descriptor of the environment, though he may recognize that the truth of a model at a particular time offers no assurance that the same model will prove true at another time or, for that matter, for another data point. Even a single verification should increase the relative worth of this specific model over those other models that could be generated on a lesser foundation.

After each iteration the investigator may review the models that were true at the time they were executed. It is natural to wonder whether a set of models contains additional information on the environment, this information residing in their similarity or trend. With this in mind, the investigator may proceed to exercise the scientific method at a higher level of abstraction. That is, he may choose to generate a model of those models that had previously proven true, a model consistent with as many of the previously successful models as possible. The new model (generally expressed within an alphabet of symbols different from those at the lower level of abstraction) can be specialized to yield another, as yet untested, model at this level of abstraction. Hopefully, this model is more likely to be successful in that it may embody some overall property of the environment that can only be found by viewing this range of experience.

In this way, the investigator can rise to higher and higher levels of abstraction. Each new level is reached by forming a model of those models that proved true at the last level of abstraction. Models of the directly observed data lie at the first level of abstraction. Models of the sequence of those models that proved true lie at the second level, models of the sequence of these that have proven true lie at a third level of abstraction, and so forth. Note that models of a higher level of abstraction must be successively reduced to first-level models in order for their validity to be tested against the observable environment. In general, fewer models comprise the set of available evidence at higher levels of abstraction. As a result the models that are generated are less likely to be verified. Increasing the level of abstraction provides the advantage of an "overview," but this benefit is reduced by the increased probability of error in modeling. The additional error associated with operating at a higher level of abstraction is balanced by the higher payoff associated with success. Verification of a high-level model may obviate costly step-by-step exploration of the intervening portion of the environment.

Scientific investigation need not be restricted to a single level of abstraction. It is often appropriate for the investigator to operate at various levels and, as new evidence accrues, to reevaluate the relative worth of operating at each level in terms of its apparent expectation of success. At each point in time, he may turn his attention to that level which offers greatest immediate promise, or he may weight all levels in terms of their relative worth, in order to formulate some combined estimate of the unknown.

Evolutionary processes allow an iterative convergence on a goal. They exist only within a domain of sufficiency with respect to that goal. New models are generated, evaluated, and used to improve previously held models only when their evaluation indicates that their retention is likely to help meet the challenge of the environment. In biological terms, each newly created organism, generated by previously "successful" creatures, is matched against its environment. Only those that exceed minimum requirements for survival can supersede their progenitors. Such successive selection contributes to the survival of the species. Each living creature may be viewed as a tentative model of some significant aspects of its environment. The domain of this evolutionary exploration is dependent on the nature of the reproductive process and the diversity contributed through natural mutation.

Social groups are formed in a similar manner, as a result of common needs in the continual struggle against an antagonistic environment. They are organized of individuals who have thus far proved to be successful at the personal level. Their structure reflects the character of these individuals and of the logic of their environment.[6] The family includes the individual, the community includes the family, the state includes the community, and so forth. At each level the structure evolves so as to reflect the nature of demonstrated success at all lower levels of abstraction with respect to the chosen goal and the properties of the encountered environment.

The efficiency of pure trial-and-error exploration is sharply reduced with an increase in the dimensionality of the domain being explored. As long as the investigator is interested only in a single aspect of his environment, pure random exploration may prove to be worthwhile. However, as soon as he attempts to map a domain of more practical interest, he encounters so many possibilities that only carefully guided trial-and-error exploration is likely to prove profitable. The advantage to be gained from employing the scientific method becomes self-evident. In man's initial exploration of the unknown, the scientific method would have been a luxury; but

---

[6]All environments may be viewed as antagonistic, in that the sheer passage of time tends to disorganize and thus threaten the survival of the decision maker.

1.7 SOME IMPLICATIONS

with the increased scope and depth of today's inquiry, using the scientific method becomes an absolute necessity.[7]

The scientific method consists of induction, inductive inference, followed by independent verification. Hypotheses, generated so that they "cover" the available evidence and some additional data points, are individually evaluated in terms of the validity of their inference. Those that prove worthy are modified, extended, or combined to form new hypotheses that carry on a "heredity of reasonableness." As the hypotheses correspond more and more closely with the logic of the environment, they provide an "understanding" that is demonstrated in terms of improved goal-seeking behavior in the face of that environment.

The correspondence between natural evolution and the scientific method is obvious. Individual organisms in nature serve as hypotheses concerning the logical properties of their environment. Their behavior is then an inductive inference about some as yet unknown aspects of their environment. Their survival is a demonstration of their suitability. Their offspring include the heredity of reasonableness as well as additional information resulting from mutation and recombination.

The evolutionary technique described above is then a realization of the scientific method in which the hypotheses are restricted only in the sense that they be finite state machines. The creation of successive hypotheses requires the introduction of randomness as well as the prevailing logic of inheritance. The versatility that is essential to intellect is a natural product of such iteration. In essence, the scientific method is a fundamental part of nature. It is no wonder, then, that its overt exercise has provided mankind with distinct benefits and now permits even its own automation through the artificial evolution of automata.

Viewed in this light, the scientific method was not *invented,* it was *discovered.* It existed long before man; in fact it gave rise to man.

As indicated, some environmental aspects of evolution can be simulated; such simulation constitutes a mechanization of the scientific method. The successive generation of models—the process of induction—can be reduced to a routine procedure. If "creativity" and "imagination" are requisite attributes of this process, these too can be realized. According to the definition offered at the beginning of this discussion, intelligence is "the ability of any decision-making entity to achieve a degree of success in seeking a wide variety of goals under a wide variety of environments." The scientific method provides an appropriate foundation for just such decision-making. In principle, the nature of the goal and the nature of the environment are in no way restricted. In a real sense, mechanization of the scientific method provides a basis for the realization of artificial intelligence. Here is a way to realize Turing's dream: "We may hope that machines will eventually compete with men in all purely intellectual fields" (Turing, 1956).

---

[7]In a curious misinterpretation, Lenat (1983) used this paragraph from Fogel et al. (1966) to suggest that I had asserted the evolutionary programming experiments were not successful, had failed because they used random variation, and that my recommendation was to incorporate heuristics and knowledge on how to modify flowcharts (i.e., finite state machines). Of course, this is in direct opposition to what I actually asserted in Fogel et al. (1966).

1.7.2 Evolutionary Problem Solving

The first generation of computer technology was devoted to the development of equipment that would carry out a large number of simple logical operations quickly and reliably. The second generation of computer technology was devoted to the development of languages that permit the detailed instruction of these equipment in an efficient manner. Computer technology is now entering a new generation, one in which it will no longer be necessary to specify exactly how the problem is to be solved. Instead, it will only be necessary to provide an exact statement of the problem in terms of a "goal" and the "allowable expenditure," in order to allow the evolution of a best program by the available computation facility. Solution of the problem may include a statement of the discovered algorithm or, if requested, the entire history of the evaluated algorithms. The old saying that "the computer never knows more than the programmer" is no longer true.

It would of course be inappropriate to utilize evolutionary programming (or other evolutionary algorithms) for the solution of problems that have already been solved. There is always an extra price to pay for versatility. That additional fee should be offered only in situations that justify the search for a new method of solution—a new logic for decision-making. At this point, we should turn to economic details, but unfortunately, at the present time little is known about such details. The programs that were used for conducting the experiments described earlier were inefficient in many respects. They were written to demonstrate feasibility rather than efficiency. Evolutionary programming should prove applicable to any digital computer of a reasonable size. The particularities of the program and the related cost of operation may be expected to depend on the capabilities of the equipment and the available programming language. Further studies in this regard would be useful.

1.7.3 A Bound to Intellect

At this point, it is natural to inquire, Is there an upper bound to intellect? Beyond the obvious constraints imposed by the sensory and motor capabilities, are there any fundamental aspects of information processing that limit the level of intelligence of a given entity? In terms of evolutionary programming there are of course physical properties of the equipment that limit what can be achieved. The speed of computation limits the number of offspring that can be evaluated within the time available for decision. The size of the memory limits the specificity of the individual models as well as the number of models that can be retained at each point in time. It is certainly possible to increase the available memory capacity, but this increase may be accompanied by a considerable increase in information access and retrieval time. This again limits the number of evaluated offspring. Clearly these factors are strongly interdependent, the best compromise being a function of the particular hardware to be used for the data processing.

Future computers will always be faster than presently available equipment. They will permit parallel data processing with obvious advantages. Such sophisticated machines will no doubt take the form of geographically removed data-processing elements that are tied together by a central control station. A decentralized computer operation permits local data processing to be carried out as a separate function. The central control is called upon only for problems that require additional capability or for problems that affect the entire system under its cognizance. A new physical limitation is introduced in such dispersed decentralized computer systems, in that the time of information transfer from element to element may become a significant factor. The capability of the entire decision-making entity may be restricted by the sheer point-to-point limitations of the physical embodiment. A number of extinct creatures may have met their demise as a result of just such internal "transmission delays." Clarke (1962) suggested that the limitation of the intelligence of a galactic-sized decision-making entity would be limited by the speed of light which imposes unavoidable transmission delays. It is easy to envisage other limitations caused by the size of the decision-making entity. Indeed, being too large may be as bad as being too small.

It is of greater interest to consider some of the logical limitations on the intelligence of a goal-seeking entity. As indicated above, intelligent behavior may be expected only if the decision-making entity possesses some suitable model of its environment. The degree of suitability is measured by the level of correspondence between the model and its real-world counterpart. Intelligence is limited if the model is too coarse, in which case it fails to offer a sufficiently precise description of the environment, or if the model is an incomplete portrayal of the relationships among the elements.

These points are worthy of clarification. The sensory capability of an entity provides some maximum resolution both in time and the number of available descriptor symbols. The better this resolution, the better the available information base. The speed of response and the specificity of the modeling process further constrain the representation of the environment. Models can only be developed within the available alphabet of relations. Inadequacy of this set of descriptors limits the accuracy of the resulting representation of the environment. For example, restricting models to a specific class of finite state machines forbids perfect representation of environments that fall outside this domain. If models can take the form of any finite state machine, they may still prove inadequate for perfect representation of some primitive recursive environments. Restricting the generality of the modeling process must be justified in terms of the adequacy of representation over the anticipated class of environments. For example, in 1962 fundamental research was conducted to find a more complete specification for the relationships that exist in the real world (Gunther, 1962). It seems that the progress made in this area is uncertain.

Then too, the intelligence of an entity may be restricted in terms of the parameters of its modeling process. For each situation there are most suitable values in the evolutionary program for the multiplicity of mutation, the weighting of recall, the available levels of abstraction, and so forth. In addition some selection must be made of an operating point between extremes of "efficiency" and "security."[8] If the environment is presumed to be almost independent of the observer, an increase of efficiency may well be justified. If, as is the usual case, the environment proves to be an intelligent adversary, it may be more appropriate to forsake efficiency for the sake of security. The more intelligent player need not be the more efficient player.

1.7.4 Evolution and Goals

Only goal-seeking entities can be intelligent. Thus far the goals have been described in terms of the relative worth (or penalty) of each possible outcome of the decisions to be made. At a more fundamental level the goal of an entity is a description of some desired future status of the system, which contains that entity and its environment. In the experiments reported above, the goal was specified to the entity by edict. In this sense the *desire* was that of the experimenter and not that of the entity. Several significant problems are raised. What is meant by the desire of an inanimate entity? How can an entity construct a model of a preferred future status? How can qualitative statements of goal be translated into quantitative expressions suitable for exact computation?

To be useful, the goal must be represented in some way within the entity. "Desirability" of any particular goal presumes the existence of a decision-making capability that can select a certain model from a set of possible models as being of greater value. This "value judgment" is a reflection of some higher-level goal that, in turn, must be selected on the basis of a still higher goal. This infinite regress may be avoided by allowing the entity an ability to model itself. As indicated previously, it should be possible to design a mechanism that would exhibit self-awareness. Such a mechanism could describe the essential features of its own information processing if so requested. With such a capability the entity may exhibit a natural tendency toward self-preservation, this being an extension of its physical inertia. Self-preservation is the highest goal in whatever complex hierarchy of subgoals the entity may hold, since all subgoals are contingent on survival. The set of possible futures is generated by combining the results of a variety of possible events. That future status is considered best which offers the highest expectation for preserving the self as portrayed by the available model. Again, the procedure is evolutionary.

The specificity of models at various levels may be very different. At higher levels the goal may be an expression of gross intent. The model of the self may be unspecific. How specific must this representation be in order to "qualify" as a valid goal or provide a satisfactory demonstration of self-awareness? Here again, these limitations depend directly on the physical embodiment of the entity in terms of its capability for information processing and storage. Just as there are degrees of intelligence, there are degrees of self-awareness.

If sufficient sensory-motor resolution is possible, the entity may be able to model individual aspects of its environment separately in terms of their observed behav-

---

[8] A similar compromise occurs in aircraft design when an operating point must be selected in the range from extreme maneuverability to complete stability.

ior. It may be able to generate a presumed goal for each observed entity and compare this goal to its own goal. Recognition of similarities and differences in goals provides a basis for social behavior. Any modification in the behavior of interacting entities establishes the existence of an organization that in itself, constitutes a new entity. The tighter this interaction, the stronger the organization. Its goal reflects the goal of its primary constituents. In recognition of its existence, it may take on a specific embodiment. Its goal gains overt expression usually in an attempt to more nearly reflect the weighted objectives of all its members. The family, the company, the community, and the nation are examples of just such organizations.

The subgoals of this organization must be generated in light of its goal and those of its members. With demonstration of successful stimulus-response behavior, the organization gains in strength and stature. It soon becomes important for each member to weigh more heavily those subgoals that support the goal of the organization to the greatest extent. Allegiance is established. The members can no longer afford to view themselves as being autonomous. The essential similarity between intelligent entities (if nothing more, their intellect) tends to make their goals similar. Thus they engage in competition. Only a higher level of social allegiance can ameliorate the resulting antagonism.

Viewing other entities in terms of their presumed goal(s) can be dangerous. Any finite sample of observed behavior can always be "explained" in terms of any one of an infinite number of goals (including "no goal," the behavior of the entity being entirely random). The selection of one of these goals as a descriptor of the observed entity appears to remove uncertainty from the situation, but this may introduce a cost in terms of security. Often the policy is to presume that the observed entity is of equal intelligence to that of the observer. The question becomes: If I had acted in that manner, what would have been my goal? This is a conservative policy, but there is always the chance that the observed entity may have greater intelligence and thus take advantage of this supposition as the basis for a costly interchange.

There is an even more fundamental danger in inferring "goals" in order to rationalize an observed behavior. Some orderliness can be found in any finite sample drawn from a perfectly random environments. This orderliness is a reflection of the observer's own characteristics. In the more general case, the environment is unknown and may be orderly to some extent. Is it possible for the observer to separately identify that portion of the observed orderliness which is a reflection of its own characteristics from that which is indigenous to the observed environment? The answer to this question hinges on the ability of the observer to "know" itself.

In order to contain the model of itself, some portion of the entity must be devoted to this purpose. This portion of the entity is not described within the model. It is, of course, possible to have a section of this portion set aside in order to model the remainder of the portion, but, here again, the representation cannot be complete. This ultimate limit also places a limit on the amount of orderliness that can be identified as a reflection of the observer. Discrimination of the orderliness of the observed environment is further degraded since each stimulus of the entity affects the observer to some degree, thus modifying its "interpretation" of successive stimuli and its model of self.

Intelligence is to some extent self-limiting. As the intelligent decision-maker gradually gains control over its environment, the severity of the remaining task is reduced. Only new goals or a radical change of environment will exercise the full intellectual capacity. The greatest development of intellect requires its continual exercise against a suitably antagonistic environment.

1.8 SUMMARY

At this point, it is appropriate to view the complete argument in perspective. Intelligence was defined in terms of the ability of a decision-maker to achieve some success in undertaking a wide variety of goals in a wide variety of environments. Versatility was recognized to be the essential ingredient of intelligent behavior. This chapter has focused on the possibility of simulating evolution as a more general means to creating an artificial intelligence.

To make the discussion specific, intelligent behavior was viewed as a composite of an ability to predict the sensed environment, coupled with an ability to select a suitable response in the light of the prediction and the given goal. The problem of predicting each next symbol was then reduced to the problem of finding that finite state machine which would have been best for achieving this goal had it existed during the available experience. This search was accomplished through a fast-time simulation of evolution in which finite state machines played the role of the evolving organisms. Random mutation of parent machines yielded offspring. These machines were driven by the available history and evaluated in terms of the given goal. Those machines demonstrating greater worth were selected to serve as the new parents. This mutation and selection was iterated with real-time decisions at any point being based on the logic of the surviving machine. The efficiency of this could be improved in a number of ways, as for example, by introducing a cost-for-complexity weighting on each machine.

But the goal need not be restricted to the prediction of each next symbol. In fact, this same evolutionary procedure is suitable for seeking any well-defined goal within the constraints imposed by the allowable expenditure. Thus, the evaluation may take place in terms of the response behavior, prediction becoming an implicit intervening variable.

A variety of prediction, classification, and control experiments were conducted, and various aspects of competitive goal-seeking were considered. Simulated evolution may offer a means for approaching the general situation in which the goals of the players may take arbitrary form and the players interact through a common environment. Such a player may model his opponent and in this manner attempt to achieve the desired control over the developing situation.

But there are even more important implications from gaining artificial intelligence through simulated evolution. The scientific method may be viewed as an evolutionary process in which a succession of models are generated and evaluated. The scientific method affords a versatile procedure for gaining a new and deeper understanding of the nature of the observed environment. Therefore, simulation of the evolutionary process is tantamount to a mechanization of the scientific method. Induction—that process which was presumed to require "creativity" and "imagination"—has been reduced to a routine procedure. Here is a technique that offers an approach to the automation of intellect, relieving the investigator so that he can turn attention to higher-level problems of selecting suitable goals and identifying new parameters of the environment that are worthy of measurement.

As indicated, the scientific method may be operated at various levels of abstraction. Each higher level operates on the environment created by the models of the lower level that had demonstrated validity. At the same time, it should be possible to devise mechanisms that will generate their own goals by referring to their internally held model of self and the "desire" for self-preservation. Such an entity might exhibit self-awareness, in that it could describe essential features of its own information transduction if so requested.

With increased sensory-motor resolution comes an ability to separately model individual aspects of the environment, to recognize the existence of similar entities, to interact with these, and, thereby, to establish a social organization. This organization in turn assumes the role of a decision-making entity with respect to its environment. Indeed, it may exhibit the same kind of goal-seeking behavior that is characteristic of its members. Their choice of goals must then take place within the context of various levels of allegiance.

Memory is built on memory. All that is sensed is lost, except for those items that can be associated with past experience in terms of identical, or at least similar, factors. All of these items are linked by the bond of the common value scale in terms of demonstrated worth of previous elicited response. Thus, without any goal, memory is blind and even survival hinges on pure chance.

And what are goals made of? They are made of the factors that contribute to self-preservation, the invariance of identity of the entity as it sees itself. Only those creatures that can effectively model themselves can alter their subgoals in support of their own survival. To survive, the self-image must be in close correspondence with the reality of themselves and their environment.

Here lies the chronicle of life. With this knowledge comes a new power to enhance memory, to manipulate goals, and to make better decisions. Of even greater significance is the possibility of creating inanimate mechanisms that possess these same qualities.

CHAPTER 2

DIVERSIFICATION

There were other independent studies that simulated evolution in the decade and a half that followed the publication of Fogel et al. (1966). These included *evolutionsstrategie* ("evolution strategies") originated by Ingo Rechenberg, Hans-Paul Schwefel, and Peter Bienert as graduate students at the Technical University of Berlin in 1964, as well as efforts to simulate genetic systems by Alex Fraser and colleagues in Australia, by Hans Bremermann at the University of California at Berkeley, and by John Holland at the University of Michigan in the United States.[1] It may be of historical interest to note, however, that the majority of efforts in evolutionary computation from the mid-1960s through 1980 appear to have been pursued within the paradigm of evolutionary programming. Principally among these undertakings were investigations by my colleagues at Decision Science, Inc. in San Diego, California, and by students of Don Dearholt at New Mexico State University. In the 1980s, and into the early 1990s, evolutionary programming diversified to include the evolution of arbitrary data structures to address a variety of function optimization problems and there were attempts to compare these approaches with other methods. Some of these efforts are reviewed below.

2.1 TWO-PERSON GAMING AGAINST NONMINIMAX PLAYERS

Although the mathematical game theory of von Neumann and Morgenstern (1947) handled the case of minimax players, comparatively little progress had been made in

---

[1]Other notable contributors included Nils Barricelli, Michael Conrad, and Woody Bledsoe. A short list cannot do justice to the vast number of individuals involved in the study of simulated evolution in the 1950s and 1960s. Readers interested in learning more about the early history of evolutionary computation should see Fogel (1998a).

2.1 TWO-PERSON GAMING AGAINST NONMINIMAX PLAYERS

	b1	b2
a1	a(1,1)	a(1,2)
a2	a(2,1)	a(2,2)

Figure 2.1 A two-by-two payoff matrix. Player A chooses the row, while player B chooses the column. The resulting entry is the payoff assigned to both players.

the development of strategies to handle nonminimax plays and exploiting errors in play[2] (see Luce and Raiffa, 1957, pp. 77–81). Burgin (1969) reported on results from evolving finite state machines to play two-person zero-sum games where the adversary was not a minimax player. He studied the following problem: Suppose two players A and B are to play a game that is defined by an $m \times n$ table of payoffs which is known to both players (see Fig. 2.1). Player A chooses the row and player B chooses the column, with the resultant entry being the payoff to both players. Assume that player B, who has the goal of minimizing his payoff, is a nonminimax player. Find a method for player A, who has the goal of maximizing his payoff, to obtain an average payoff greater than the "value" of the game. In other words, player A must learn from the experience gained over the course of moves in the game by recognizing patterns in player B's actions and react to this information in a useful manner.

Burgin (1969) used evolutionary programming on finite state machines where each input symbol represented the pair of previous moves made by players A and B (Fig. 2.2). Using a population of finite state machines mutated by a variety of possible alterations of the number of states and transitions between states, Burgin evolved these representations of the behavior of player A against an algorithm for player B where the $(\mu + 1)$st move was determined as

$$j_{\mu+1} = j_{\mu+1 \text{ opt}},$$

where

$$E j_{\mu+1 \text{ opt}} = \sum_{i=1}^{m} \xi_i a_{ij} \leq E_l = \sum_{i=1}^{m} \xi_i a_{il}$$

---

[2]Morgenstern (1966) commented ". . . the theory is unable to deal with mistakes made by the opponents, an issue that was mentioned originally in the book by von Neumann and myself. The indication was, at least at that time, that one really would have to have fundamentally new ideas in order to deal with the situation."

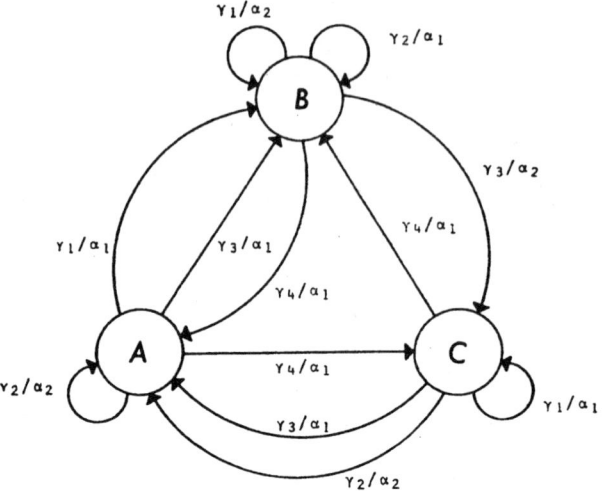

Figure 2.2 A finite state machine representing a strategy in the two-player game. The input symbols γ_1–γ_4 represent the four possible pairs of moves between players A and B, given that each player has two alternative moves. The output symbols α_1 and α_2 are player A's response given the current input symbol (from Burgin, 1969).

for all $l = 1, \ldots, n$. The values ξ_i satisfy:

$$\sum_{i=1}^{m} \xi_i = 1$$

The value for $j_{\mu+1 \text{ opt}}$ was made unique by posing that if several values of j yielded the same optimal expected payoff, the smallest j would be selected. Note that the above algorithm will defeat an opponent playing a mixed random strategy with an invariant probability distribution that is not equal to the proper minimax distribution.

The 4 × 4 payoff matrix in Figure 2.3 was used for the evolutionary program-

$$A \begin{array}{c} \\ \end{array} \begin{array}{|cccc|} \hline 2 & 3 & 1 & 4 \\ 1 & 2 & 3 & 4 \\ 2 & 3 & 4 & 1 \\ 4 & 1 & 2 & 2 \\ \hline \end{array}$$

with B above the matrix.

Figure 2.3 The four-by-four payoff matrix used in Burgin (1969). Each entry is the payoff to both players A and B, where player A chooses the row and player B chooses the column. Note that each row contains a 4, and each column contains a 1, so the best possible payoff is 1 for the minimizing player and 4 for the maximizing player.

ming experiments. The minimax value of the game was 2.4375, and the best possible score for a maximizing player on any move was 4, while for the minimizing player it was 1. The minimax solutions for player A is to play each of the four moves with probabilities {16/48, 3/48, 17/48, 12/48} and for player B: {5/16, 3/16, 4/16, 4/16}. The evolutionary program was run as a maximizing player against the algorithm in three separate trials (see Fig. 2.4), and to eliminate any biases, also as minimizing player against the algorithm and as minimizing player against a proper minimax player. Each game was played for 500 moves. From Figure 2.4 the evolutionary program clearly learned to take advantage of the nonminimax algorithm and achieved payoffs that were superior to the value of the game. Thus the evolutionary procedure offered an approach toward generalized gaming, where adversaries may interact in zero-sum or non-zero-sum games with time-varying strategies. Later efforts in these regards were offered by Fogel (1991a) and others using the iterated prisoner's dilemma (see Chapter 3).

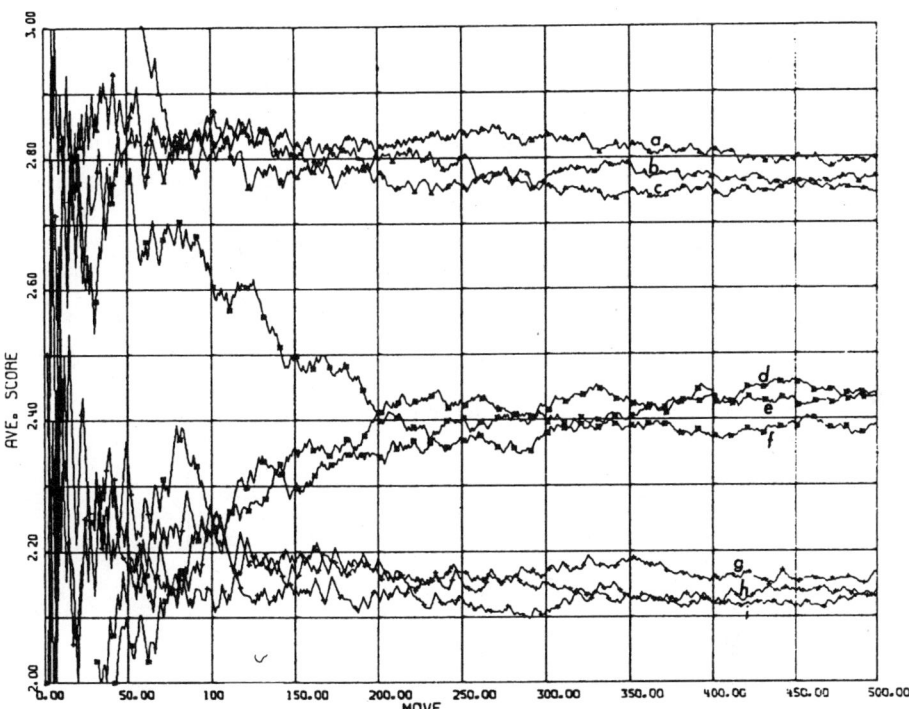

Figure 2.4 The average payoff in two-person zero-sum games of the evolutionary program against a nonminimax playing computer algorithm. Curves (a), (b), and (c) indicate the evolving maximizing player against the algorithm, curves (g), (h), and (j) indicate the evolving minimizing player against the algorithm, and curves (d), (e), and (f) indicate using evolutionary programming to play the minimizing player against a minimax algorithm. The value of the game is 2.4375 (from Burgin, 1969).

2.2 COEVOLUTION

Rather than play the evolutionary program against a fixed algorithm, it was of interest to observe the effects of having one evolutionary strategy play against another coevolving strategy. Fogel and Burgin (1969) reported a series of gaming experiments conducted over a three-year period that extended the games mentioned above; in one of these experiments, the performance of an evolutionary program was made a function of how well it interacted with an opposing evolutionary program. Attention was given to the same payoff matrix as shown in Figure 2.3, with each player given a recall length of 100 symbols and imposed penalties for complexity of 0.03 and 0.05 per state for each of two parent finite state machines in each population. Over 500 moves, team 1 evolved a 10-state machine, while team 2 evolved a 6-state machine. The final score was 2.412 (recall that the minimax value of the game was 2.4375), which favored team 1 as the minimizing team. Additional experiments were conducted by varying the recall length between 20 and 100 symbols (shorter recall was favored), and overall, average performance was seen to converge toward the minimax solution.

Of greater interest, however, was using evolutionary programming to learn strategies in non-zero-sum games, as often encountered in combat situations. The payoff functions are described in the form of Figure 2.5 where the payoffs to player A is a_{ij} when A selects the ith row and B selects the jth column. Similarly b_{ij} represents the payoff to player B. If $a_{ij} = -b_{ij}$ for all i and j then the game is zero sum, but in general, this need not be the case. A variety of game forms can be described by varying the values of a_{ij} and b_{ij}. Figure 2.6 shows three games: (1) the *coordination* game, in which both players receive the maximum return when playing in the first row and column, and have no incentive to move otherwise, (2) the *trust* game (also

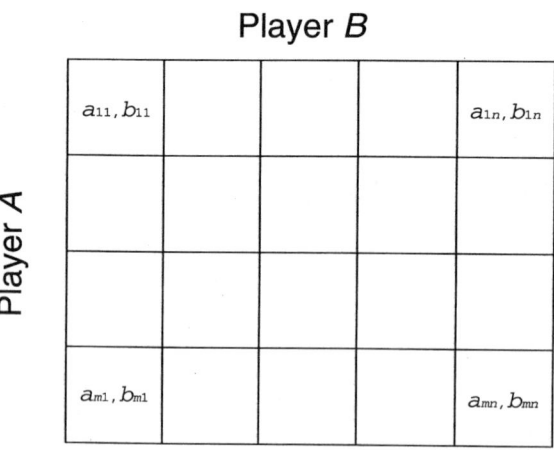

Figure 2.5 The general payoff matrix where player A chooses a row and player B chooses a column. The payoff is indicated as a pair, where the former value is awarded to player A and the latter is awarded to player B.

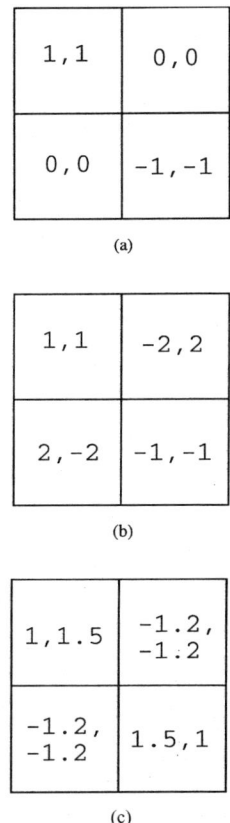

Figure 2.6 Payoff matrices for the (*a*) coordination game, (*b*) trust (or prisoner's dilemma) game, and (*c*) bargaining game.

commonly referred to as a *prisoner's dilemma*), where both players can profit by moving in the first row and column, but each can make a still larger profit by defecting into the second row or column while the opponent does not defect, and (3) the *bargaining* game, where more than one cell yields a maximum point of return, but the return is divided equally between the two players. Every non-zero-sum game has at least one equilibrium point (Nash, 1954).

Considering the equilibrium point to be a solution to the game can be paradoxical. For example, the game described in Figure 2.7 has the equilibrium point (a_2, b_2); however, this is the only point where both players lose. In fact a game may have more than one equilibrium point,[3] neither of which may be viewed as a "solution" in

---

[3]The equilibrium point in a two-person non-zero-sum game has similar properties to the saddlepoint in a two-person zero-sum game: the point that describes to both players an alternative such that if either player individual deviates from his best strategy, he will be penalized.

5,5	-10,10
10,-10	-5,-5

Figure 2.7 The payoff matrix has a single equilibrium point (a_2, b_2), but this is only point where both players lose.

the usual sense of the word. For example, the prisoner's dilemma game shown in Figure 2.8 has equilibrium points (a_1, b_2), (a_2, b_1).

For coevolutionary instances of the coordination game, the evolving finite state machines converged very rapidly to having both sides play the first row and column (Fogel and Burgin, 1969). In the trust game, the evolution converged on an equilibrium, but which equilibrium could not be known a priori. In four experiments with the payoff matrix shown in Figure 2.8, the strategies converged in three runs to (a_1, b_2) after 12, 27, and 3 moves, and (a_2, b_1) in the fourth run after 12 moves. This was in agreement with the expectation of Luce and Raiffa (1957, p. 105): "It is our impression that players of a game do, in some sense evolve to an equilibrium position, not necessarily a unique one—and we can say that, from a descriptive point of view, the set of equilibrium points of a game do constitute a characterization of the solution of the game." In bargaining games, the evolutionary programs also converged to one of the equilibrium points.

2.3 PURSUIT AND EVASION

Pursuit-evasion games are of particular interest, not only because of their practical importance, but because they represent a class of games that cannot be described in normal form and that are not known to have a generally applicable analytic solution. In place of an analytic formulation, such games require repetitive simulation in order to determine the expected result. This requires objective definition of the rules of the

1,1	-2,2
2,-2	-5,-5

Figure 2.8 The prisoner's dilemma payoff matrix used in evolutionary programming experiments from Fogel and Burgin (1969).

game, the number of teams taking part in the game, the allowed moves at each point in time (thus reflecting the dynamic and kinematic constraints of motion), the objectives of each team or player, the penalties imposed for certain actions (e.g., a team may be required to lose a player), the amount of information given to the player concerning the motion of the opponent, and the termination rules of the game.

An evolutionary program can be made to play against an algorithm as either pursuer or evader, or it can be used to simulate two teams or players who play against each other, one representing the pursuer and the other representing the evader. The worth of the evolved strategies (e.g., in the form of finite state machines) may be based on the payoff matrix used in measuring their performance over the recall (the fit-score, i.e., how well the finite state machine fit the available data) or through the use of another payoff matrix for evaluating the action of the player. These payoff matrices may depict the cost as a direct function of the distance between pursuer and evader (e.g., in terms of an all-or-none condition, where a move in the right direction means no error while a move in any wrong direction corresponds to an error of some fixed cost), or as a function of the position and direction of the move of the opponent. If more than two players are involved, the payoff matrices are defined in a similar manner but are naturally of greater size.

Fogel and Burgin (1969) conducted a series of evolutionary pursuit-evasion games of increasing levels of realism. In the first of these, a 500×500 rectilinear grid served as the domain of the game. The evader was permitted to move one unit in the $\pm x$ or $\pm y$ direction, while the pursuer could move one or two units with the same freedom. The evader was programmed to move along a 45° line by alternating x and y moves. Experiments resulted in a finite state machine that used the last three moves of the evader to predict its next location, allowing the pursuer to catch the target in 272 moves. The second experiment put the evader in a quarter-circle, and the evolutionary program again caught the evader, this time requiring 206 moves but with a longer recall length (20 moves).

A further experiment required the pursuing evolutionary program to reduce the distance between its position and the evader's position four moves in advance. To obtain this prediction, the finite state machines were driven with the input symbol obtained as a result of the evader's play four moves into the past. The best machine was then driven by the sequence up to the present move so that its output would correspond to a prediction of the evader's fourth move into the future. This determined whether or not it was possible for the pursuer to reach the predicted position of the evader within four moves, given constraints. If so, the algorithm determined the next four moves of the pursuer; if not, the next move of the pursuer was determined so as to minimize the distance between the predicted position of the evader and the position of the pursuer after the next move. This look ahead served to accelerate the rate of convergence of the pursuer on the evader (Fogel and Burgin, 1969).

To make the pursuit-evasion games more realistic, the evader was given a variety of tasks, each having different priority, and noise was imposed on the game so the pursuer no longer obtained perfect information about the direction and/or position of the opponent. In all cases the evolutionary program was used to aid the pursuer. However, clear advantage in this regard could also be turned to aid the evader.

More specifically, the evader also served as an attacker. In the initial phase of the game, the evader's goal was to move toward a specific target (the attack). The pursuer was assumed to predict the present position and motion of the attacker (evader) and move accordingly. When the pursuer reached a certain distance from the attacker (the detection radius), the attacker modified his goal in order to avoid the pursuer by maximizing the distance between the two. If, however, the pursuer reached a still closer radius (the lock-on range), the evader modified his goal once again so that he moved in a maximally irregular pattern, thus reducing his probability of being killed. The pursuer could not fire his weapon beyond a certain range, and a kill was presumed if the pursuer had correctly predicted the evader's next position (within this range) before firing his weapon.

Both the pursuer and evader operated under dynamic constraints. The evader-attacker could move in two-unit steps in the semiquadrant direction so long as he did not change his direction from that of the previous move. In each move, he could choose to change direction by ± 45°. If he changed direction, however, he could

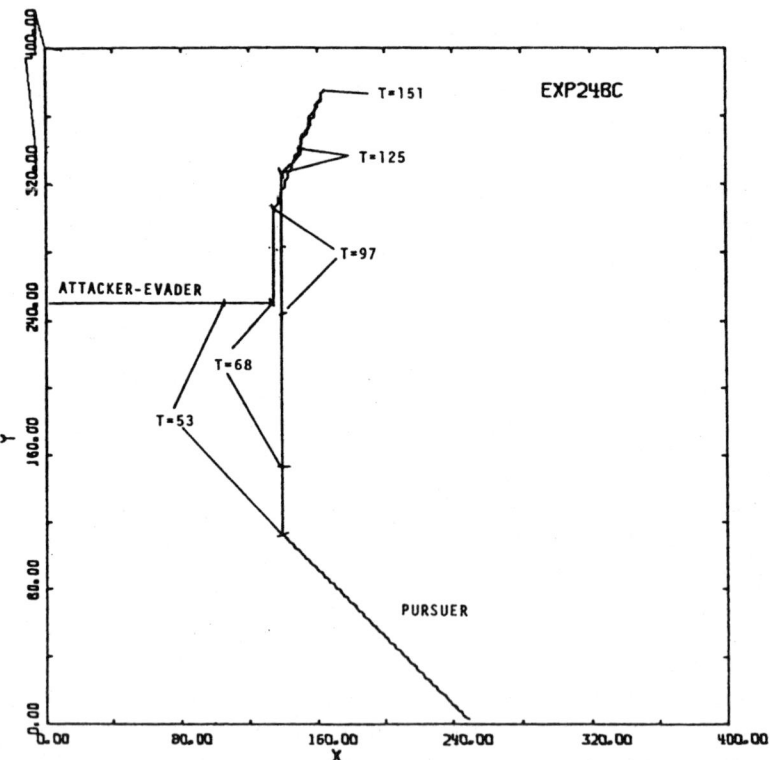

Figure 2.9 The course of action between the pursuer and the attacker-evader in the noiseless environment. The pursuer starts at (250, 0) and the attacker-evader starts at (0, 250) with the target goal of reaching the line $x = 500$. The graph indicates that the pursuer was able to successfully intercept the attacker-evader (from Fogel and Burgin, 1969).

move only one-unit length in the next step. The same constraints applied to the pursuer, except that he could move in three-unit steps if there were no change of direction and one-unit step immediately after each change of direction. The pursuer behaved according to an algorithm using the evolutionary program to predict the motion of the evader. The best finite state machine evolved by the program, which represented the evader's strategy, was driven four steps into the future. The next four moves of the pursuer were then determined to be those that would minimize the distance to the pursuer's projected position. This number of moves was arbitrary. Clearly it could be adjusted as a function of distance between the players, the fit-score of the finite state machine representing the evader's strategy, or other criteria.

A considerable number of experiments were performed with an evader's detection range of 100 units, a lock-on range of 70 units, and a weapon range of 4 units. The evader started at the Cartesian position (0, 250) with its target being the line $x = 500$. The pursuer started at (250, 0). Figure 2.9 shows the course of action in a noiseless environment, while Figure 2.10 shows a pursuit in a noisy environ-

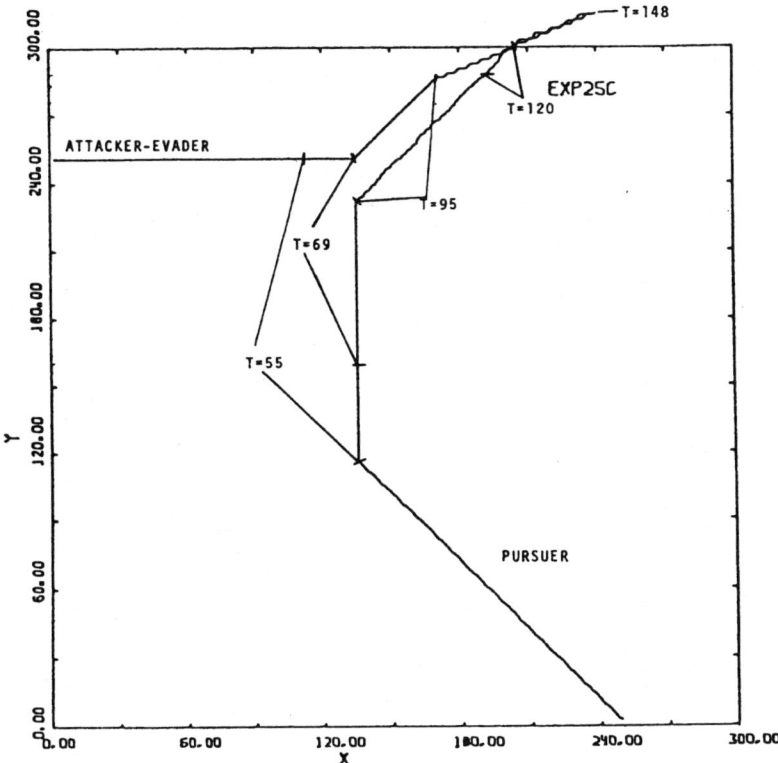

Figure 2.10 The course of action between the pursuer and the attacker-evader when unbiased noise is imposed on the position of the attacker-evader. The pursuit in noisy conditions required greater distance to intercept than observed in Figure 2.9 (from Fogel and Burgin, 1969).

56 DIVERSIFICATION

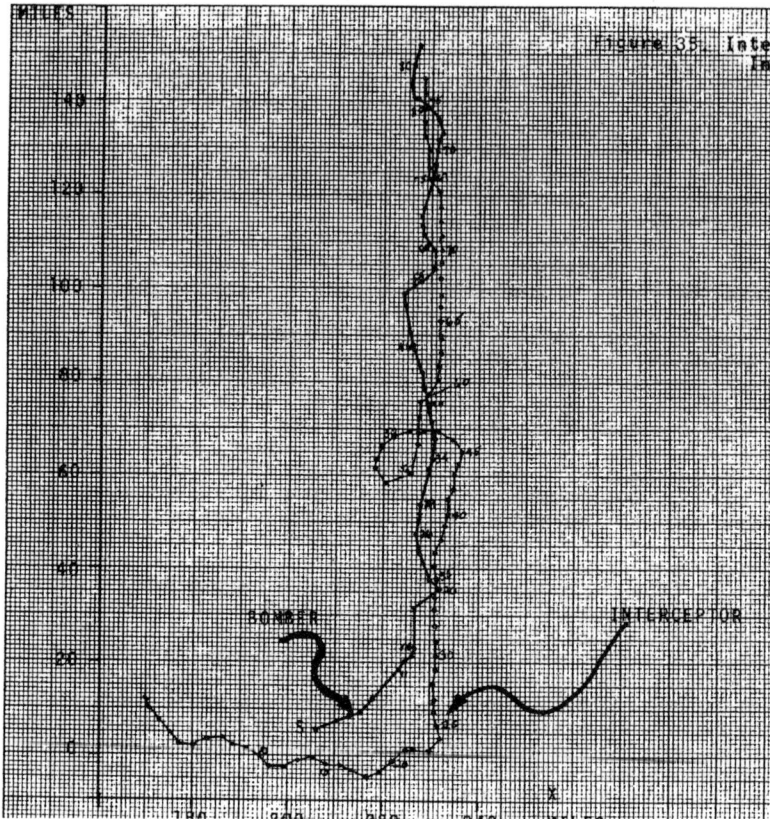

Figure 2.11 The evolved path of an interceptor pursuing and catching a bomber using associated flight dynamics. The simulated engagement covered almost 120 miles (from Fogel and Burgin, 1969).

ment. Additional experiments were conducted where the evader's behavior during final closure was determined by a finite state machine whose behavior was predicted by the evolving pursuer (coevolution). Evolutionary simulations of interceptors pursuing bomber aircraft (with associated flight dynamics in three dimensions and a nine-symbol coding of control stick position) were also undertaken Figure 2.11.

2.4 MODELING TIME SERIES

Following the initial success in evolving finite machines to induce sequences of symbols (Fogel et al., 1966; and others), Fogel and Moore (1968) used an evolutionary program to model a linear dynamic system, a nonlinear dynamic system, and a human operator in single-axis tracking tasks. The previous evolutionary program

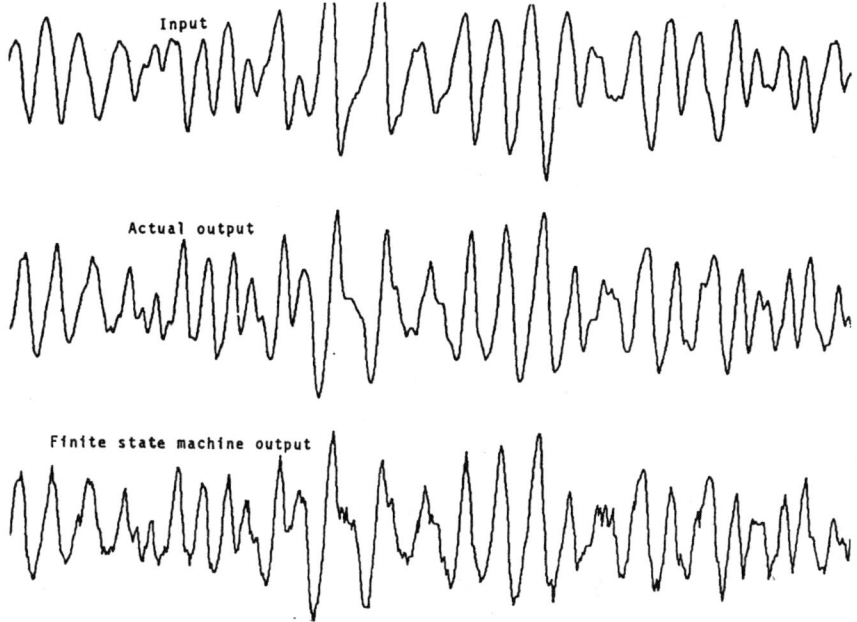

Figure 2.12 A typical result from Fogel and Moore (1968) where finite state machines were evolved to identify input-output response of a linear dynamic system. The upper time series is the input sequence, the middle time series is the output of the system. The lower time series is the predicted output best evolved finite state machine having 63 states. The qualitative (visual) goodness-of-fit to the observed data appears good.

(from Fogel et al., 1966) was extended to handle 64 input symbols and a maximum of 32 states per machine. A linear dynamic system of the form

$$\frac{k_1 A + k_1 k_2 s}{(A + s)^2}$$

was studied as a common instance of a model used to describe a human operator in tracking control, the constants being chosen to typify human control of a second-order plant having the transfer characteristic k/s^2 through the use of compensatory tracking. A random disturbance was used to drive this plant with the input and output data (350 points) quantized into 64 equal class intervals of amplitude.[4] Evolutionary programming was used to find finite state machines that would minimize the magnitude of the difference between their output and the plant's output given the series of input data (see Fig. 2.12 for one example from a series of related experiments).

---

[4] As a historical aside, these points were transcribed to punch cards at a rate of five points per second.

58 DIVERSIFICATION

For the nonlinear system, the attempt was to choose a system comprising linear segments, its logic being: When the motion of the plant is away from the reference, the analog "human operator" responds in proportion to the rate of this motion. If the plant is not moving or is moving toward the reference, there is no response unless the rate is such that a simple prediction indicates that the plant would cross the reference in less than T seconds, this causing the operator to respond in proportion to the rate of motion. The prediction is of the form

$$x(t + T) = x(T) + T\dot{x}(T),$$

where $x(T)$ is the present displacement of the plant from the reference and $\dot{x}(T)$ is the present rate of the plant displacement. All outputs of the system were transformed by a second-order approximation to a transport delay and by a lag filter to represent the inertia of the "pilot's" musculature. The input and output of the nonlinear system were recorded while it controlled a second-order plant in order to track a random variable. The sampling rate was increased to 100 samples per second. Figure 2.13 indicates the agreement between an evolved finite state machine and the output data over a recall of 4000 samples. Figure 2.14 shows the results of evolving finite state machines to model a human operator controlling the same second-order plant.

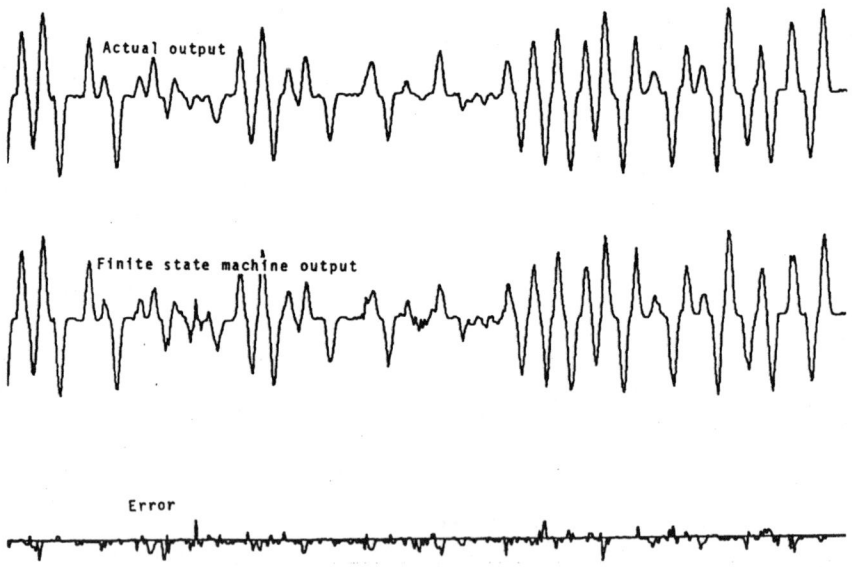

Figure 2.13 A typical result from Fogel and Moore (1968) where finite state machines were evolved to predict the output of a nonlinear system (see text for description of the system). The upper and middle time series are the actual and predicted output using a finite state machine having 63 states. The lower time series shows the prediction error. Evolution was conducted using a "magnitude of the difference" scoring (i.e., absolute error).

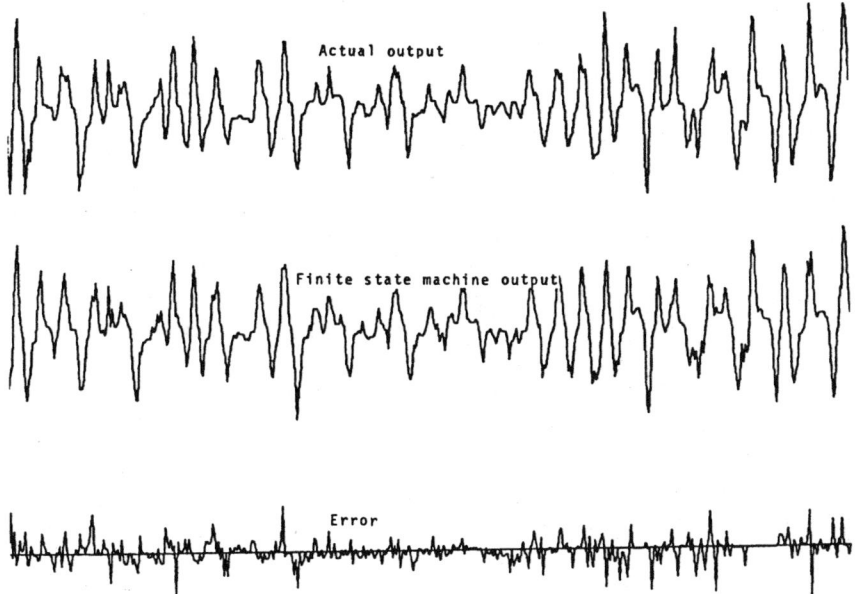

Figure 2.14 A typical result from Fogel and Moore (1968) where finite state machines were evolved to fit the output of a second-order nonlinear system based on the input of a human operator. The error appeared slightly higher than observed in Figure 2.13.

Other efforts in evolutionary modeling of time series included Lutter and Huntsinger (1969), who examined a simple evolutionary programming procedure using a single parent and offspring to predict the water temperature of a simulated cooling tower (Lutter and O'Connor, 1967) based on actual hourly temperature and dew point readings taken over two weeks in January 1968 in Rapid City, South Dakota (also see Lutter, 1968), and Burgin (1974) who evolved models of flight data from an X-15.

2.5 PATTERN RECOGNITION

The greatest amount of evolutionary programming research in the 1970s was conducted at New Mexico State University under the guidance of Don Dearholt. Much of this work (e.g., Root, 1970; Cornett, 1972; Lyle, 1972; Cornett et al., 1973; Holmes, 1973; Trellue, 1973a, 1973b, 1974; Montez, 1974; Anderson and Dearholt, 1974; Vincent, 1976; Williams, 1977; and others) remained unknown to the evolutionary programming community until it was uncovered by David Fogel in 1997 while preparing Fogel (1998a). The majority of these efforts involved the evolution of finite state machines (Moore machines) to perform pattern recognition of regular expressions and handwritten characters. Several novel additions to the basic framework of evolutionary programming were offered and some are described here.

Lyle (1972) offered the possibility of a variable number of parents and offspring, with adaptable probabilities for mutation operators (suggesting explicitly that there could be an ideal setting), as well as the number of times a parent was to be mutated in generating an offspring. With a problem of inducing regular expressions in Moore machines, Lyle offered alternative scoring functions using a ratio of the distance to the final state after the input sequence is applied to the maximum distance from any point to any other point on the graph (i.e., the machine). Distance was defined as the number of transitions required to reach the final state. This served as an alternative to including explicit penalties for complexity based on the number of states of the machines evolved.

Holmes (1973), following Lyle (1972), incorporated a mechanism for dynamically adjusting mutation probabilities, as well as the number of mutations to apply, in evolving finite state machines to classify integers coded in binary digits as being prime or nonprime. Using populations of five parents, numeric values were set controlling the relative frequency of performing a given mutation operator. The likelihood of increasing the probability of the occurrence of a given mutation or of increasing the number of mutations was given by the weight of that component in the dynamic weight vector minus the weight of the preceding component (the components were ordered), divided by the value of the last component of the vector. Additional features were included to reinforce changes that appeared to generate improved solutions. It appears that this dynamic mutation strategy was self-adaptive at the level of the population (i.e., there was one set of mutation parameters controlling the random variation to be applied to all parents in the population which was periodically adjusted on the basis of how well the population was performing). Machines were evolved to recognize primes and nonprimes coded as binary numbers of up to eight digits and then tested on integers up to 250 (see Table 2.1).

Trellue (1973b), following Cornett (1972), studied the evolution of finite state machines to identify handprinted characters. The objective was to allow the evolutionary program to decide which features of coded characters to use to assist in discriminating between various characters. Input strings corresponded to the digitization of the input pattern. (For greyscale, this would require a nonbinary alphabet.) Classification was determined by the final state of the evolved finite state machines (as many as 75 states were possible). A five-parent population was used with one offspring per parent and user-defined weights on the probabilities of possible modes of mutation (these were later optimized off-line). The number of mutations per offspring was varied, with two and three giving the best results. Four different types of grids for digitizing (Fig. 2.15) six variations of the letters A, B, C, D into binary strings to serve as input to the evolving machines (Fig. 2.16 depicts the training set). Figure 2.17 shows learning rates for two of the four digitization schemes.

Montez (1974) used evolutionary programming to classify different types of

Table 2.1 Results from five experiments with evolutionary programming to predict the primeness of the increasing integers encoded in binary strings

Experiment	Program	Training Set Interval	Evolved at Generation	States	Training Set Score with Lower Bound 3			Score on Interval 3 to 250		
					Primes Correct/Total Primes	Nonprimes Correct/Total Nonprimes	Totals	Primes Correct/Total Primes	Nonprimes Correct/Total Nonprimes	Totals
1	APL	0–15	21	17	5/5	6/8	11/13	51/52	126/196	177/248
2	APL	16–31	30	14	5/5	8/11	13/16	36/52	148/196	184/248
3	APL	3–31	80	12	10/10	14/19	24/29	35/52	153/196	188/248
4	ALGOL	3–31	70	19	9/10	19/19	28/29	19/52	178/196	197/248
5	ALGOL	3–100	680	15	16/24	73/74	89/98	23/52	187/196	210/248

Source: Holmes (1973).

Note: The columns indicate the type of programming language, the training interval, the generation number at which the machine in question was evolved, the size of the evolved machine, and the performance on the training set (see the third column for the range of the training interval) and the primes from 3 to 250.

62 DIVERSIFICATION

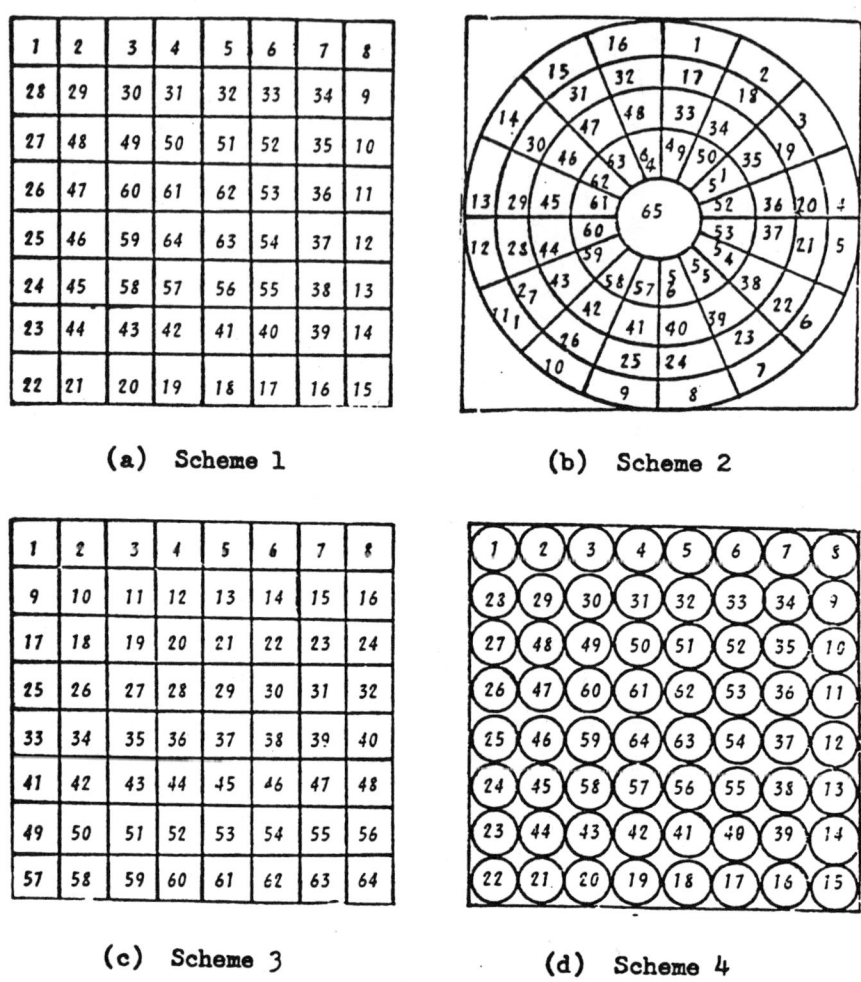

Figure 2.15 The four different digitization schemes used in Trellue (1973b) for scanning handwritten characters A, B, C, D.

electrocardiograms. Training sets consisted of digitized strings of different electrocardiograms (Fig. 2.18) as characterized by specific features of interest (Montez, 1974). In all, 46 electrocardiograms were selected and digitized, with the evolutionary program classifying 33 of these correctly (normal vs. abnormal) after 270 generations (Fig. 2.19) (also Dearholt, 1976; Vincent, 1976;[5] Williams, 1977).

---

[5]Vincent (1976) looked at both electrocardiogram and handwritten character classification. Moreover he demonstrated both mathematically and empirically that earlier criticism of evolutionary programming in Lindsay (1968) was misplaced. Lindsay (had) denounced the evolutionary programming results in Fogel et al. (1966) as being no better than a completely random search of all possible finite state machines. Unfortunately, the analysis used to come to this conclusion was in error (for details, see Fogel, 1992a).

2.5 PATTERN RECOGNITION   63

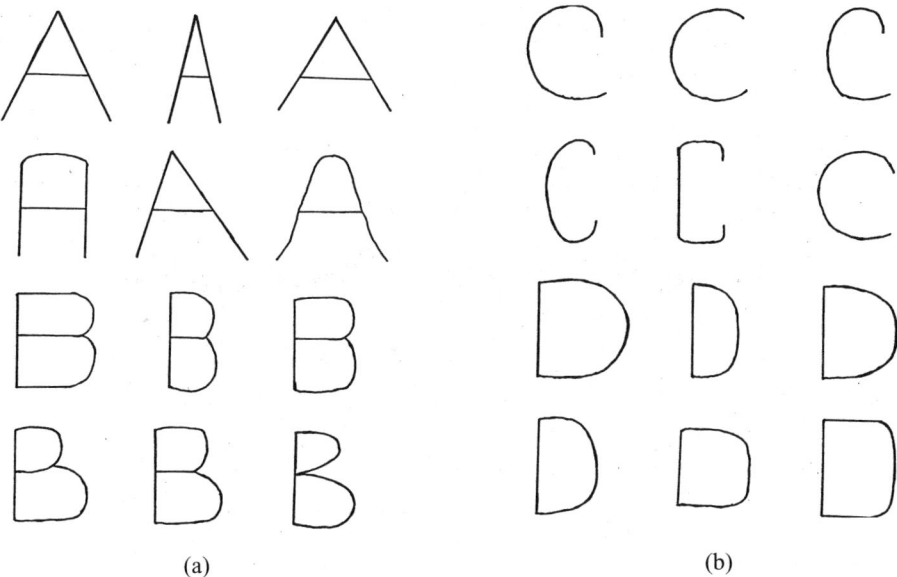

Figure 2.16 The set of letters that were scanned into binary strings and used for training in Trellue (1973b). Six variations of (*a*) letters A and B, (*b*) letters C and D.

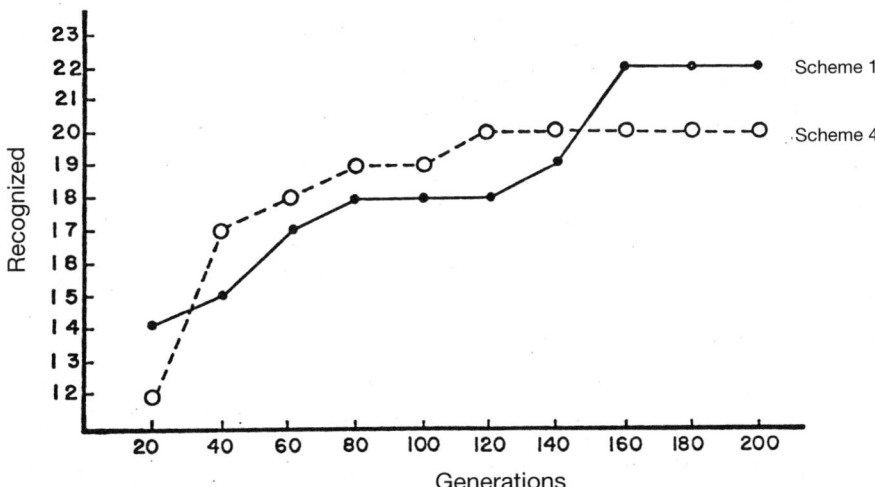

Figure 2.17 The learning curves from Trellue (1973b) for scannng scheme 1 (dark circles) and scheme 4 (open circles) in recognizing the 24 patterns of binary strings taken from the six variations of A, B, C, and D. The curves have a shape that is typical of evolutionary optimization where there is rapid initial learning followed by asymptotic convergence to an optimum. For this particular trial, none of the machines in the initial population of 10 were able to recognize any strings successfully (all were initialized to a default machine).

64 DIVERSIFICATION

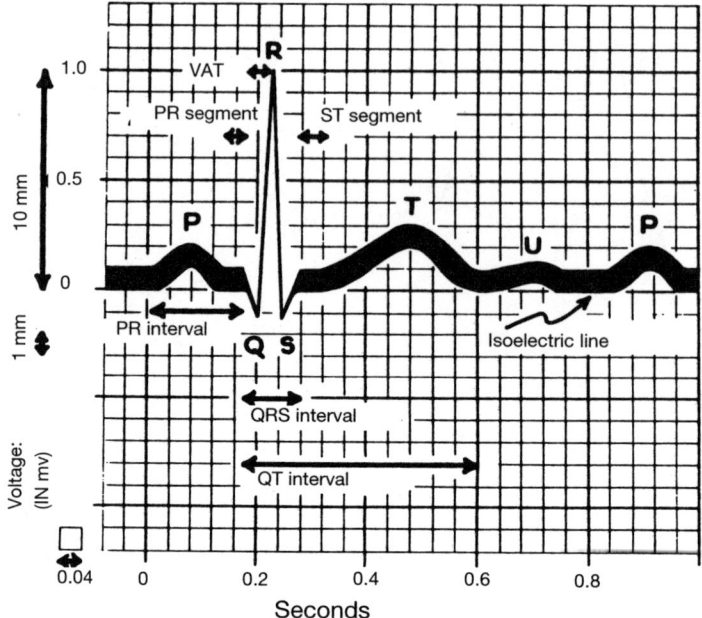

Figure 2.18 Quantizing an electrocardiogram (ECG) (from Montez, 1974). Interest was given to classifying whether or not the digitized ECG represented normal or abnormal conditions.

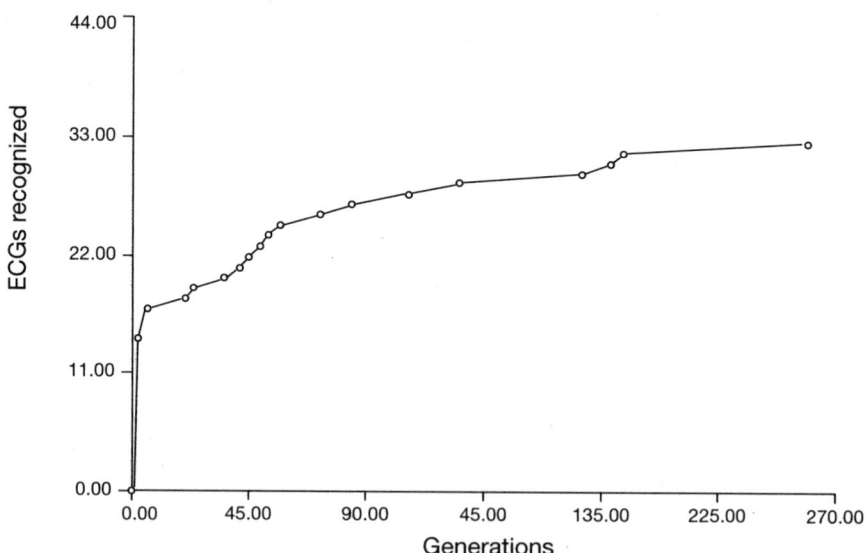

Figure 2.19 Results from Montez (1974) evolving finite state machines to recognize normal and abnormal ECGs. Again, the rate of initial learning was rapid, followed an asymptotic approach to an optimum. Here, 33 ECGs out of 46 were classified correctly in 270 generations.

2.6 SIMULATED ECOSYSTEMS AND THE NATURE OF INTELLIGENCE

Atmar (1976), in his doctoral dissertation completed under the guidance of Don Dearholt at New Mexico State University, sought primarily to identify the fundamental aspects of intelligent organisms. Atmar defined intelligence as "that property in which the organism senses, reacts to, learns from, and subsequently adapts its behavior to its present environment in order to better promote its own survival." By this definition, there are three organizational forms of intelligence: (1) phylogenetic (intelligence arising within the evolving line of descent), (2) ontogenetic (arising within the individual), and (3) sociogenetic (arising with the group). These three forms are essentially equivalent in process; each includes a reservoir of learned behavior and a unit of mutability. However, ontogenetic and sociogenetic intelligence are learned "tricks" of phylogenetic intelligence that serve to increase an organism's ability to predict its environment, thereby minimizing surprise.

Atmar (1976) conducted a simulation of evolving finite state machines (following Fogel et al., 1966) whereby machines were placed in an artificial environment under rules of engagement that defined an ecosystem. The environment was divided into a grid of cells. Each cell was defined by an environmental pattern of 0s and 1s (a repeating string). Finite state machines were tested in their ability to predict the sequence in their current cell; those that failed to meet minimal prediction criteria were removed from the population. Competition for survival occurred when five or more machines entered the same cell. The possibility was also incorporated for survival to be based on trophic levels in which larger machines required the presence of smaller machines (as a predator requires prey) and had to predict the requisite environment better than any machine of a lower trophic level (simulating a general observation that predators are more complex and better adapted than their prey). Surviving machines had the possibility to migrate to neighboring cells (migration was in proportion to the size of the machines, with larger machines being able to migrate to more distant cells than smaller machines). Figure 2.20 indicates a series of epochs of such a simulation.

Atmar (1976) also examined the use of evolving finite state machines for pattern recognition of handwritten characters (with special attention to patterns not recognized to belong to a particular class) and discussed the realization of evolutionary machines in hardware, a topic that has received recent attention (Sipper et al., 1997; and others).

2.7 SEQUENCE INDUCTION WITH DETERMINISTIC AUTOMATA

Tomita (1982) used a simplified evolutionary programming procedure to construct a deterministic finite automaton that would accept examples of strings from a "right-list" and reject strings in from a "wrong-list." In particular, a population consisting of only a single parent that generated a single offspring by mutation was used. The deterministic finite automata took the form of a machine with a finite number of states. Some of these states were designated as "final" states, with transi-

<EPOCH 1>

```
0 0 0 0 0 0 0 0 0 0 0 0 0 0 0 0 0 0 0 0
0 0 0 0 0 0 0 0 0 0 0 0 0 0 0 0 0 0 0 0
0 0 0 0 0 0 0 0 0 0 0 0 0 0 0 0 0 0 0 0
0 0 0 0 0 0 0 0 0 0 0 0 0 0 0 0 0 0 0 0
0 0 0 0 0 0 0 0 0 0 0 0 0 0 0 0 0 0 0 0
0 0 0 0 0 0 0 0 0 0 0 0 0 0 0 0 0 0 0 0
0 0 0 0 0 0 0 0 0 0 0 0 0 0 0 0 0 0 0 0
0 0 0 0 0 0 0 0 0 0 0 0 0 0 0 0 0 0 0 0
0 0 0 0 0 0 0 0 0 1 0 0 0 0 0 0 0 0 0 0
0 0 0 0 0 0 0 0 1 1 1 0 0 0 0 0 0 0 0 0
0 0 0 0 0 0 0 0 0 2 0 0 0 0 0 0 0 0 0 0
0 0 0 0 0 0 0 0 0 0 0 0 0 0 0 0 0 0 0 0
0 0 0 0 0 0 0 0 0 0 0 0 0 0 0 0 0 0 0 0
0 0 0 0 0 0 0 0 0 0 0 0 0 0 0 0 0 0 0 0
0 0 0 0 0 0 0 0 0 0 0 0 0 0 0 0 0 0 0 0
0 0 0 0 0 0 0 0 0 0 0 0 0 0 0 0 0 0 0 0
0 0 0 0 0 0 0 0 0 0 0 0 0 0 0 0 0 0 0 0
0 0 0 0 0 0 0 0 0 0 0 0 0 0 0 0 0 0 0 0
0 0 0 0 0 0 0 0 0 0 0 0 0 0 0 0 0 0 0 0
```

(a)

<EPOCH 5>

```
0  0  0  0  0  0  0  0  0  0  0  0  0  0  0  0  0  0  0  0
0  0  0  0  0  0  0  0  0  0  0  0  0  0  0  0  0  0  0  0
0  0  0  0  0  0  0  0  0  0  0  0  0  0  0  0  0  0  0  0
0  0  0  0  0  0  0  0  0  0  0  0  0  0  0  0  0  0  0  0
0  0  0  0  0  0  0  0  0  49 0  0  0  0  0  0  0  0  0  0
0  0  0  0  0  0  0  0  1  4  50 0  0  0  0  0  0  0  0  0
0  0  0  0  0  0  0  4  52 28 28 28 0  0  0  0  0  0  0  0
0  0  0  0  0  0  51 58 59 33 28 35 40 0  0  0  0  0  0  0
0  0  0  0  0  36 57 39 60 33 33 62 65 63 0  0  0  0  0  0
0  0  0  0  36 36 39 39 39 33 48 18 40 19 1  0  0  0  0  0
0  0  0  0  0  36 64 39 18 48 48 23 65 2  0  0  0  0  0  0
0  0  0  0  0  0  46 46 47 43 48 48 23 0  0  0  0  0  0  0
0  0  0  0  0  0  0  46 47 48 48 23 0  0  0  0  0  0  0  0
0  0  0  0  0  0  0  0  47 47 23 0  0  0  0  0  0  0  0  0
0  0  0  0  0  0  0  0  7  0  0  0  0  0  0  0  0  0  0  0
0  0  0  0  0  0  0  0  0  0  0  0  0  0  0  0  0  0  0  0
0  0  0  0  0  0  0  0  0  0  0  0  0  0  0  0  0  0  0  0
0  0  0  0  0  0  0  0  0  0  0  0  0  0  0  0  0  0  0  0
0  0  0  0  0  0  0  0  0  0  0  0  0  0  0  0  0  0  0  0
```

(b)

Figure 2.20 The results of simulating an artificial ecology in Atmar (1976). Each cell in the enviroment posed a binary cyclic prediction problem for an evolving set of finite state machines. In this case each cell posed the same prediction problem (a homogeneous environment). The simulation starts with a single finite state machine (organism 1) in the center. The simulation is displayed at epochs (a) 1, (b) 5, (c) 8, and (d) 10. The eventual dominant machine (organism 39) was created before epoch 5 (b) whereupon it spread rapidly. Epoch 8 (c) shows that the greatest diversity was apparent on the periphery of the expanding populations, where selection pressures were lowest.

2.7 SEQUENCE INDUCTION WITH DETERMINISTIC AUTOMATA

<EPOCH 8>

```
0  0  0  0  0  0  0  0  0  0  0  0  0  0  0  0  0  0  0  0
0  0  0  0  0  0  0  0  0  0  49 0  0  0  0  0  0  0  0  0
0  0  0  0  0  0  0  0  49 54 50 0  0  0  0  0  0  0  0  0
0  0  0  0  0  0  0  4  4  50 50 50 0  0  0  0  0  0  0  0
0  0  0  0  0  0  51 58 50 50 50 50 50 0  0  0  0  0  0  0
0  0  0  0  0  50 51 39 50 50 50 50 50 50 0  0  0  0  0  0
0  0  0  0  36 36 39 39 39 50 48 50 50 28 28 0  0  0  0  0
0  0  0  36 36 39 39 39 39 39 48 48 28 28 63 63 0  0  0  0
0  0  36 36 39 39 39 39 39 39 48 48 62 62 63 63 0  0  0  0
0  36 36 39 39 39 39 39 39 39 39 48 48 63 63 19 26 0  0  0
0  0  36 36 39 39 39 39 39 39 39 48 48 48 23 65 2  0  0  0
0  0  0  46 46 39 39 39 39 39 48 48 48 48 48 23 0  0  0  0
0  0  0  0  23 47 39 39 39 48 48 48 48 48 23 0  0  0  0  0
0  0  0  0  0  47 47 39 48 48 48 48 48 23 0  0  0  0  0  0
0  0  0  0  0  0  47 47 48 48 48 48 23 0  0  0  0  0  0  0
0  0  0  0  0  0  0  47 47 48 48 23 0  0  0  0  0  0  0  0
0  0  0  0  0  0  0  0  47 47 23 0  0  0  0  0  0  0  0  0
0  0  0  0  0  0  0  0  0  7  0  0  0  0  0  0  0  0  0  0
0  0  0  0  0  0  0  0  0  0  0  0  0  0  0  0  0  0  0  0
0  0  0  0  0  0  0  0  0  0  0  0  0  0  0  0  0  0  0  0
```

(c)

<EPOCH 10>

```
0  0  0  0  0  0  0  0  7  54 50 0  0  0  0  0  0  0  0  0
0  0  0  0  0  0  0  4  54 50 50 50 0  0  0  0  0  0  0  0
0  0  0  0  0  0  51 54 50 50 50 50 50 0  0  0  0  0  0  0
0  0  0  0  0  50 51 39 50 50 50 50 50 50 0  0  0  0  0  0
0  0  0  0  50 50 39 39 39 50 48 50 50 50 50 0  0  0  0  0
0  0  0  50 50 39 39 39 39 39 48 48 50 50 50 50 0  0  0  0
0  0  36 36 39 39 39 39 39 39 48 48 50 50 28 28 0  0  0  0
0  36 36 39 39 39 39 39 39 39 48 48 48 28 62 63 63 0  0  0
36 36 39 39 39 39 39 39 39 39 39 48 48 62 62 63 63 0  0  0
36 39 39 39 39 39 39 39 39 39 39 39 48 48 63 63 19 26 0  0
36 36 39 39 39 39 39 39 39 39 39 39 48 48 48 50 65 2  0  0
0  36 46 39 39 39 39 39 39 39 39 48 48 48 48 48 50 0  0  0
0  0  23 47 39 39 39 39 39 39 48 48 48 48 48 50 0  0  0  0
0  0  0  47 47 39 39 39 39 48 48 48 48 48 23 0  0  0  0  0
0  0  0  0  47 47 39 39 48 48 48 48 48 23 0  0  0  0  0  0
0  0  0  0  0  47 47 48 48 48 48 48 23 0  0  0  0  0  0  0
0  0  0  0  0  0  47 47 48 48 48 48 23 0  0  0  0  0  0  0
0  0  0  0  0  0  0  47 47 48 48 23 0  0  0  0  0  0  0  0
0  0  0  0  0  0  0  0  47 47 23 0  0  0  0  0  0  0  0  0
0  0  0  0  0  0  0  0  0  7  0  0  0  0  0  0  0  0  0  0
```

(d)

Figure 2.20 (*c, d*) (continued).

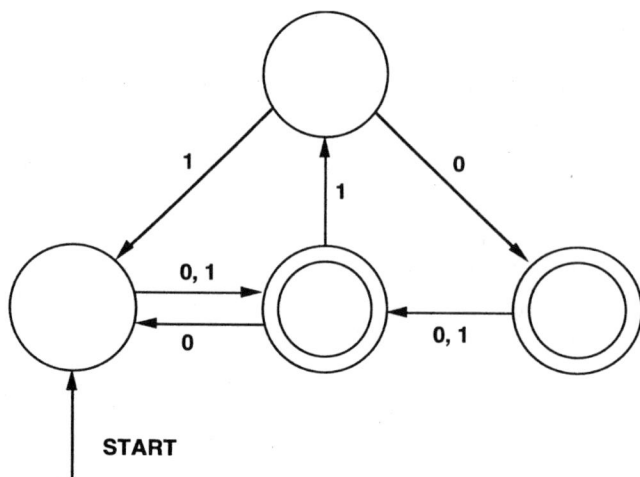

Figure 2.21 A possible deterministic finite automaton from Tomita (1982) used in evolutionary programming experiments to induce sequences of symbols. The states with two circles are "final" states. For an input string, if there was a transition from the initial state to any of the final states, then the string was said to be "accepted"; otherwise it was rejected. The problem was to evolve machines that would correctly accept and reject posed strings based on properties of similarity.

tions between states depending on the input symbol observed (see Fig. 2.21). Seven distinct binary environments were considered (see Table 2.2), along with their "inverse" problems in which the right-list was swapped with the wrong-list. Tomita (1982) demonstrated that the evolutionary approach could generate automata to solve each of the problems posed, and the results compared favorably with purely random sampling in terms of the expected number of sampled machines that would be required to solve each problem instance (Table 2.3).

2.8 REVISITING AND EXTENDING EARLY EVOLUTIONARY PROGRAMMING

In 1985 I began an effort with my son David Fogel to return to the early experiments offered in Fogel et al. (1966) and recapitulate them using the faster computers that had become available. This allowed for repeated trials, larger population size, and confidence limits to be generated on the expected behavior of evolving finite state machines in problems in prediction, identification and control. The results were documented in a series of quarterly reports culminating in the final report Fogel and Fogel (1986).

The prediction experiments included a population of size 12, alternative cyclic environments in two, four, and eight symbols (possibly corrupted by noise), as well as sequences generated by Markov transition matrices or Fibonacci series, varying

Table 2.2 Accept and reject lists for the seven problems

Problem 1

Accept List	Reject List
( )	(0)
(1)	(1 0)
(1 1)	(0 1)
(1 1 1)	(0 0)
(1 1 1 1)	(0 1 1)
(1 1 1 1 1)	(1 1 0)
(1 1 1 1 1 1)	(1 1 1 1 1 1 1 0)
(1 1 1 1 1 1 1)	(1 0 1 1 1 1 1 1)
(1 1 1 1 1 1 1 1)	

Problem 2

Accept List	Reject List
( )	(1)
(1 0)	(0)
(1 0 1 0)	(1 1)
(1 0 1 0 1 0)	(0 0)
(1 0 1 0 1 0 1 0)	(0 1)
(1 0 1 0 1 0 1 0 1 0 1 0)	(1 0 1)
	(1 0 0)
	(1 0 0 1 0 1 0)
	(1 0 1 1 0)
	(1 1 0 1 0 1 0 1 0)

Problem 3

Accept List	Reject List
( )	(1 0)
(1)	(1 0 1)
(0)	(0 1 0)
(0 1)	(1 0 1 0)
(1 1)	(1 1 1 0)
(0 0)	(1 0 1 1)
(1 0 0)	(1 0 0 0 1)
(1 1 0)	(1 1 1 0 1 0)
(1 1 1)	(1 0 0 1 0 0 0)
(0 0 0)	(1 1 1 1 1 0 0 0)
(1 0 0 1 0 0)	(0 1 1 1 0 0 1 1 0 1)
(1 1 0 0 0 0 0 1 1 1 0 0 0 0 1)	(1 1 0 1 1 1 0 0 1 1 0)
(1 1 1 1 0 1 1 0 0 0 1 0 0 1 1 1 0 0)	

Problem 4

Accept List	Reject List
( )	(0 0 0)
(1)	(1 1 0 0 0)
(0)	(0 0 0 1)
(1 0)	(0 0 0 0 0 0 0 0)
(0 1)	(1 1 1 1 1 0 0 0 0 1 1)
(0 0)	(1 1 0 1 0 1 0 0 0 0 0 1 0 1 1 1)
(1 0 0 1 0 0)	(1 0 1 0 0 1 0 0 0 1)
(0 0 1 1 1 1 1 1 0 1 0 0)	(0 0 0 0)
(0 1 0 0 1 0 0 1 0 0)	(0 0 0 0 0)
(1 1 1 0 0)	
(0 1 0)	

(continued)

Table 2.2 Accept and reject lists for the seven problems (*continued*)

Problem 5

Accept List	Reject List
( )	(1)
(1 1)	(0)
(0 0)	(1 1 1)
(1 0 0 1)	(0 1 0)
(0 1 0 1)	(0 0 0 0 0 0 0 0 0)
(1 0 1 0)	(1 0 0 0)
(1 0 0 0 1 1 1 1 0 1)	(0 1)
(1 0 0 1 1 0 0 0 0 1 1 1 1 0 1 0)	(1 0)
(1 1 1 1 1 1)	(1 1 1 0 0 1 0 1 0 0)
(0 0 0 0)	(0 1 0 1 1 1 1 1 1 1 1 0)
	(0 0 0 1)
	(0 1 1)

Problem 6

Accept List	Reject List
( )	(1)
(1 0)	(0)
(0 1)	(1 1)
(1 1 0 0)	(0 0)
(1 0 1 0 1 0)	(1 0 1)
(1 1 1)	(0 1 1)
(0 0 0 0 0 0)	(1 1 0 0 1)
(1 0 1 1 1)	(1 1 1 1)
(0 1 1 1 1 0 1 1 1 1)	(0 0 0 0 0 0 0 0)
(1 0 0 1 0 0 1 0 0)	(0 1 0 1 1 1)
	(1 0 1 1 1 1 0 1 1 1 1)
	(1 0 0 1 0 0 1 0 0 1)

Problem 7

Accept List	Reject List
( )	(1 0 1 0)
(1)	(0 0 1 1 0 0 1 1 0 0 0)
(0)	(0 1 0 1 0 1 0 1 0 1)
(1 0)	(1 0 1 1 0 1 0)
(0 1)	(1 0 1 0 1)
(1 1 1 1 1)	(0 1 0 1 0 0)
(0 0 0)	(1 0 1 0 0 1)
(0 0 1 1 0 0 1 1)	(1 0 0 1 0 0 1 1 0 1 0 1)
(0 1 0 1)	
(0 0 0 0 1 0 0 0 0 1 1 1 1)	
(0 0 1 0 0)	
(0 1 1 1 1 1 1 1 1 1 1 1)	
(0 0)	

Source: Tomita (1982).

2.8 REVISITING AND EXTENDING EARLY EVOLUTIONARY PROGRAMMING

Table 2.3 The number of steps (iterations) required to solve the seven problems and their inverses by evolutionary programming, as well as the expected number of steps required to solve each problem under blind random search

Problem	Evolution	Random Search
1	98	33
2	134	315
3	2052	> 50000
4	442	12500
5	1768	> 50000
6	277	50000
7	206	50000
~1	300	167
~2	89	1862
~3	1939	> 50000
~4	246	> 50000
~5	1844	> 50000
~6	886	> 50000
~7	3725	> 50000

Source: Tomita (1982).

Note: Inverses are denoted by the tilde (~). Although random search would be expected to discover solutions to simple problems faster than evolutionary search, the evolutionary approach was much more efficient on the more complex problems.

payoff functions (all-or-none, absolute error, squared error), and different mutation rates and number of mutations per parent. A particularly interesting case concerned the environment 101001000 . . . with a sequence of increasing number of 0s followed by a 1. With a payoff function as shown in Figure 2.22, the best-evolved machines changed their predictive behavior in light of the dynamic environment (Fig. 2.23). They first tended to predict the symbol 1, but after the sequence increased to have eight or more 0s in-a-row, the evolutionary program recognized the advantage of switching to predicting mostly 0s. Another case involved flipping all of the sym-

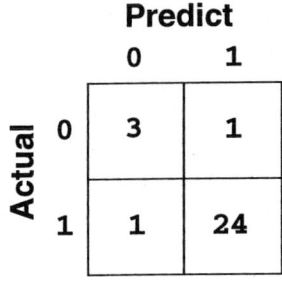

Figure 2.22 The payoff function used in Fogel and Fogel (1986) in predicting the sequence 101001000 . . . . The value of correctly predicting a 1 was made eight times greater than correctly predicting a 0, this in view of the increasing rarity of 1s in the environment. The errors were equally costly. From Fogel and Fogel (1986).

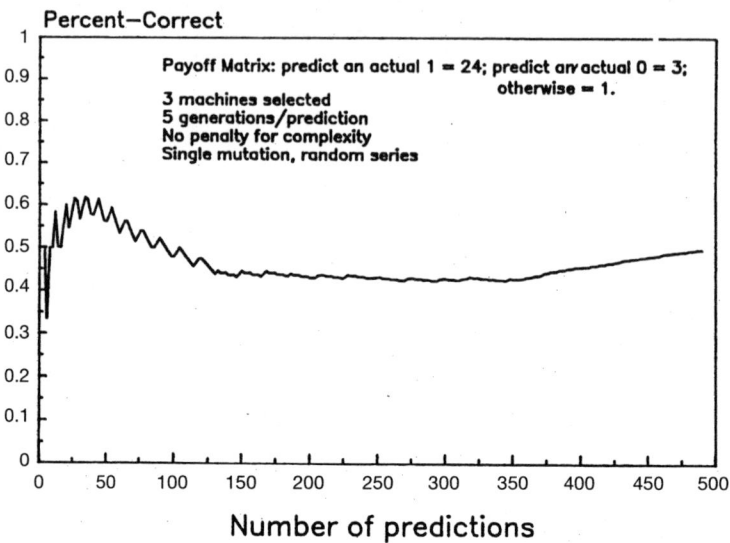

Figure 2.23 The cumulative fraction of correct predictions as a function of the number of predictions when using evolutionary programming with a population size of three FSMs to forecast a sequence with an increasing number of 0s between each 1. After a sufficient number of 0s has been observed, it is of greater benefit to predict 0 rather than 1. The asymptotic behavior of the best evolved predictor was eventually to simply predict 0 (from Fogel and Fogel, 1986).

bols in the cyclic sequence (101110011101)* and observing how long the evolutionary program took to learn the new regularity (Fig. 2.24). The evidence indicated that the prior evolution did not hinder learning the completely obverse environment (see Fig. 2.25 for a similar experiment in a four-symbol alphabet).

The second quarterly report offered some of the first comparisons of one-parent mutation operators to two-parent recombination. This followed the first international conference on genetic algorithms (Grefenstette, 1985) in which several papers stressed the importance of including crossover in evolutionary simulations (Goldberg and Lingle, 1985; Goldberg, 1985; and others).[6] For the cyclic environment (101110011101)*, experiments with both a 2-state and 10-state crossover operator (remove and replace a collection of states from one parent machine with another) showed no qualitative difference in performance as compared with the previous mutation-based approach. Experiments with the sequence 101001000 . . . using the pay-

[6]Goldberg and Lingle (1985) evolved solutions to the traveling salesman problem and wrote that "with only a unary operator to search for the better string orderings, we have little hope of finding the best ordering, or even very good orderings, in strings of any substantial length," instead promoting the use of recombination operators. Such an (uncritical) advocacy was surprising because it ignored the then well-known work of Kirkpatrick et al. (1983), published in *Science,* where annealing (using a unary operator) generated very good solutions to the traveling salesman problem.

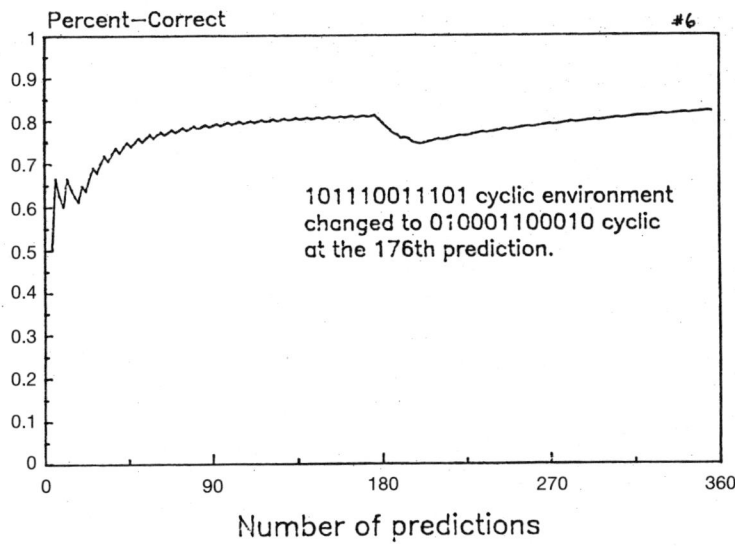

Figure 2.24 The cumulative fraction of correct predictions as a function of the number of predictions when using evolutionary programming with a population size of three FSMs to forecast a sequence that is flipped to the obverse symbols after 176 predictions. There is an initial reduction in performance when the environment shifts, but the evolutionary programming quickly learned the new environment (from Fogel and Fogel, 1986).

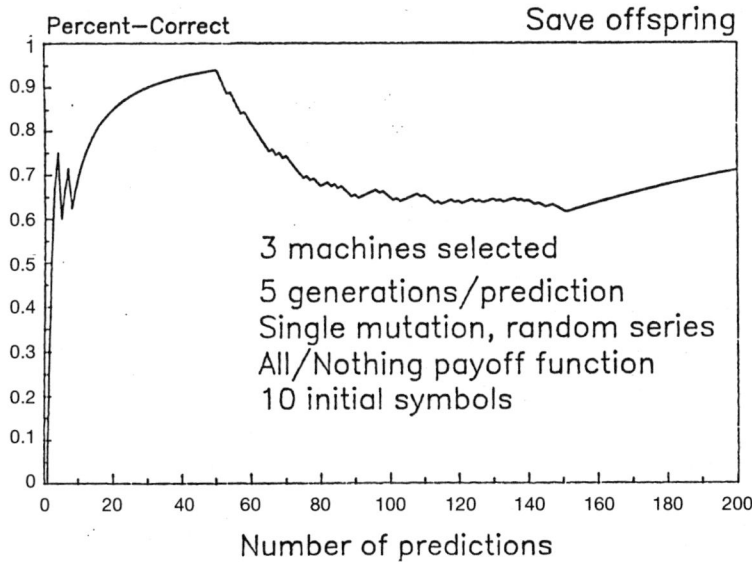

Figure 2.25 Similar to Figure 2.24, the evolutionary programming is made to predict a series of changing environments in a four-symbol alphabet. After generating a perfect predictor to the first sequence (note the asymptotic approach to 1.0), the environment is changed to (331022) for 15 repetitions. Performance then degrades and approaches a cumulative fraction correct of about 0.65. When the initial environment is repeated, the evolutionary programming quickly readapts to this sequence and again generates a perfect predictor (from Fogel and Fogel, 1986).

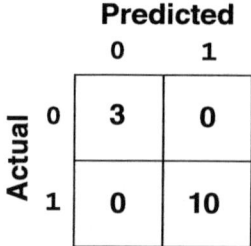

Figure 2.26 Payoff matrix used in experiments with and without crossover on evolving finite state machines to predict the sequence 101001000. . . . (from Fogel and Fogel, 1986).

off matrix in Figure 2.26 indicated the use of crossover to generate slightly worse predictive performance than omitting crossover (averaged over 15 trials). These initial trials led to more in-depth comparisons in a wider range of problems (e.g., traveling salesman problem, general function optimization, which are discussed below).

The second quarterly report also offered the first use of an order-based representation in evolutionary programming, as applied to solving the traveling salesman problem (TSP). A single parent-single offspring procedure was introduced using an ordered list of cities to represent a candidate solution. Mutation randomly removed and replaced a city in the parent tour list, thus generating an offspring tour. Experiments considered uniform TSPs (i.e., the cities were distributed uniformly at random in a specified region) of up to 256 cities, as well as a case of 90 cities grouped in 9 clusters of 10 cities. The third quarterly report extended efforts to include multiple traveling salesman and three-dimensional routing problems (Figs. 2.27 and 2.28).

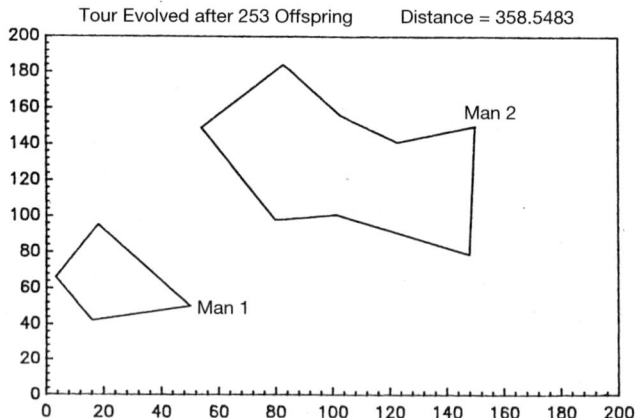

Figure 2.27 A best-evolved solution for a two-person traveling salesman problem. "Man 1" and "Man 2" indicate the starting positions for the salesmen. The objective was to have a shortest combined tour length while having the cities distributed equally to each salesman. This necessitated a trade-off which resulted in essentially an automatic clustering of cities. Only 253 solutions were examined to generate the tours shown here (from Fogel and Fogel, 1986).

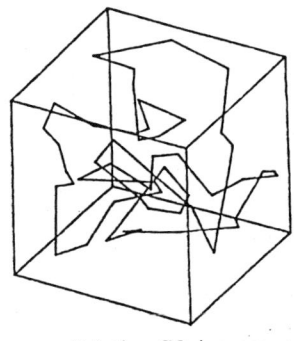
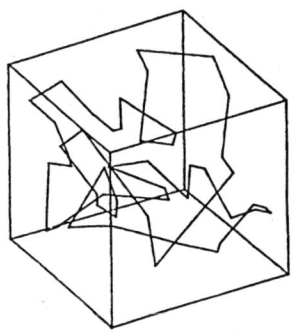

Rotation: 30 degrees   Rotation: 60 degrees

Figure 2.28 The evolutionary program was extended to handle traveling salesman problems in three dimensions (e.g., as might be suitable for aircraft or submersibles) (from Fogel and Fogel, 1986).

Efforts were also made to compare directly the use of the partially mapped crossover operator (PMX) from Goldberg and Lingle (1985) on 100-city uniform TSPs, and the mutation-based evolutionary programming approach was shown to be statistically significantly better in generating solutions with a lower total distance. These experiments led to more concentrated effort in applying evolutionary programming to routing problems, and further comparisons between the effectiveness of different variation operators in evolutionary algorithms.

2.9 ROUTING PROBLEMS

Following Fogel and Fogel (1986), Fogel (1988) offered the first population-based evolutionary programming approach to the traveling salesman problem that relied on mutation (i.e., no recombination) and selection. Experiments were conducted using 50 parents, each creating a single offspring using the remove-and-replace operator described above. Fogel (1988) also offered a probabilistic selection method, in contrast to previous efforts that retained only the best individuals. Specifically, for each individual, say X, in the population at every generation, 10 other individuals, $Y_1, \ldots, Y_{10}$, were selected from the population at random as competitors. Pairwise comparisons were conducted in which the probability of attaining a "win" in each encounter was set equal to

$$P(X \text{ receives win}) = \frac{f(Y_i)}{f(X) + f(Y_i)},$$

where Y_i is the ith competitor and the function $f(\cdot)$ returns the Euclidean distance of the individual tour. For example, if a tour with length 1000 competes with a tour of length 2000, the probability for the first tour to obtain a win would be 2000/(1000 +

2000) = 2/3. This procedure was chosen by viewing the assigned error score (i.e., the distance of the tour) as the mean time to obtain a resource under an exponential distribution. Given two exponential random variables E_1 and E_2, the probability that $E_1 < E_2$ is

$$\frac{\mu_{E_2}}{\mu_{E_1} + \mu_{E_2}},$$

where μ_{E_1} is the mean of E_1 and likewise for E_2.

One of the noteworthy results from Fogel (1988) came in addressing ten 100-city TSPs where the cities were randomly distributed with uniform likelihood. The evolutionary program evaluated 400,000 offspring in each trial and the average best tour length over the ten trials was 966.7602. A computer program was written to sample 2500 tours at random for each trial, and it found an estimated mean tour length of 5118.2538 with an estimated standard deviation of 544.6610. Using a normal approximation to the error distribution of the tour lengths (the error was taken to be the difference between the tour length and the expected best tour length and calculated using the statistical mechanics formula offered in Bonomi and Lutton (1984)), the average evolved solution was 7.6222σ below the mean. Thus, after evaluating 8.58×10^{-151} total number of tours, the mean best tour discovered by evolutionary programming was superior to 99.99999999999% of all possible tours. Compared with the results of the PMX operator on 30 100-city TSPs again indicated statistically significant evidence that the mutation-based approach generated superior solutions.

Ambati et al. (1991) applied a similar mutation-based approach using strict deterministic selection. Rather than remove and replace a randomly chosen city from an ordered list, variation was imposed by exchanging two cities. The number of exchanges to apply to each parent in creating and offspring was random following a hypergeometric distribution. Ambati et al. (1991) proved that their algorithm could achieve solutions with a running time of $O(n \log n)$, where there are n cities, that were typically less than 25% longer than the expected optimum tour. Empirical evidence indicated that the use of the graded set of variation operators (depending on the number of mutations applied to a parent) allowed for more diversity of step size and offered an advancement over the method in Fogel (1988).

Fogel (1993a) extended Ambati et al. (1991) by introducing a mutation operator that reversed the list of a tour segment instead of exchanging the two end points. The first city in the operation was selected uniformly at random and the next was selected to be a distance $[2, n/2]$ units away, again uniformly at random. This inversion operator (2-OPT) effectively removes crossing portions of circuits (see Fig. 2.29). Results on a 30-city, 50-city, and 75-city TSP indicated either a known best solution or an improvement over previous best available solutions (Fig. 2.30). When applied to a 1000-city TSP, the best-evolved tour after 40 million candidates had been evaluated was estimated to be between 5.1 and 7.4% longer than the expected best possible routing (Fig. 2.31).

Fogel (1993b) studied the number of tours required for evaluation in order to find solutions that were no worse than 10 percent longer than the expected best for TSPs

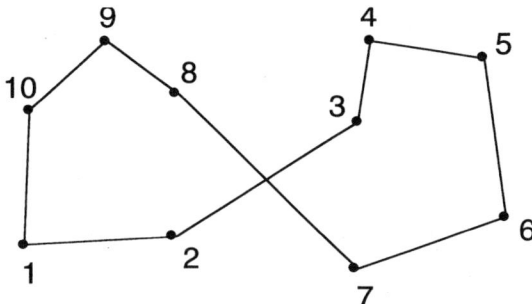

Figure 2.29 The inversion operator picks two cities along the route and reverses the path between those cities. Here the tour is [1, 2, 3, 4, 5, 6, 7, 8, 9, 10], with a return to city 1 implied at the end of the list. The tour crosses itself between paths (2, 3) and (7, 8). Applying the reversal to the pair (3, 7) generates the list [1, 2, 7, 6, 5, 4, 3, 8, 9, 10]. This removes the crossing and generates a shorter path. Note that applying the reversal to the pair (2, 8) generates the list [1, 8, 7, 6, 5, 4, 3, 2, 9, 10] which actually induces an additional crossing. In this case the reversal should not be applied to the cities between the chosen pair, but rather to the outside of the chosen pair. This would generate [9, 8, 3, 4, 5, 6, 7, 2, 1, 10] which again removes the crossing. Only the former instance of the operator was used in Fogel (1993a).

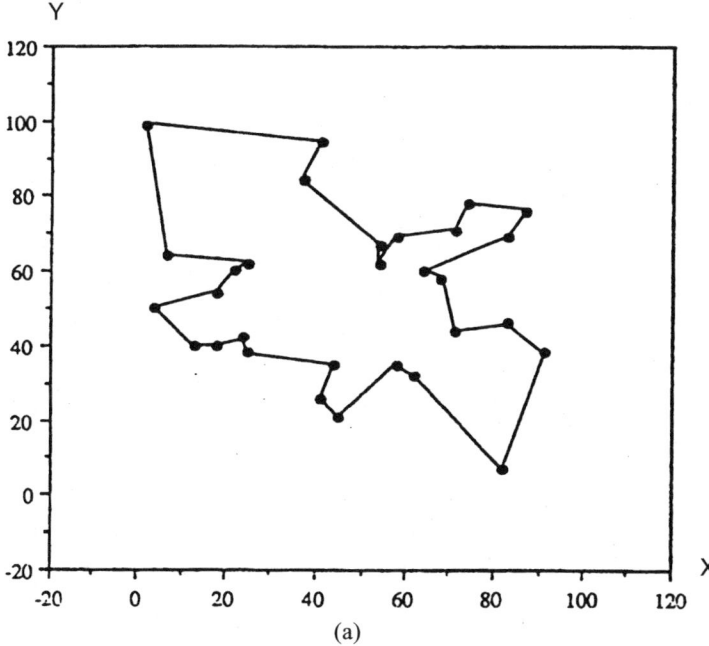

Figure 2.30 The best-evolved solutions from Fogel (1993a) on a (*a*) 30-, (*b*) 50-, and (*c*) 75-city problem from the literature. For the case of 30 cities, this is the known best possible tour. The solutions for the other two problems were better than the best available solutions when published.

(continued)

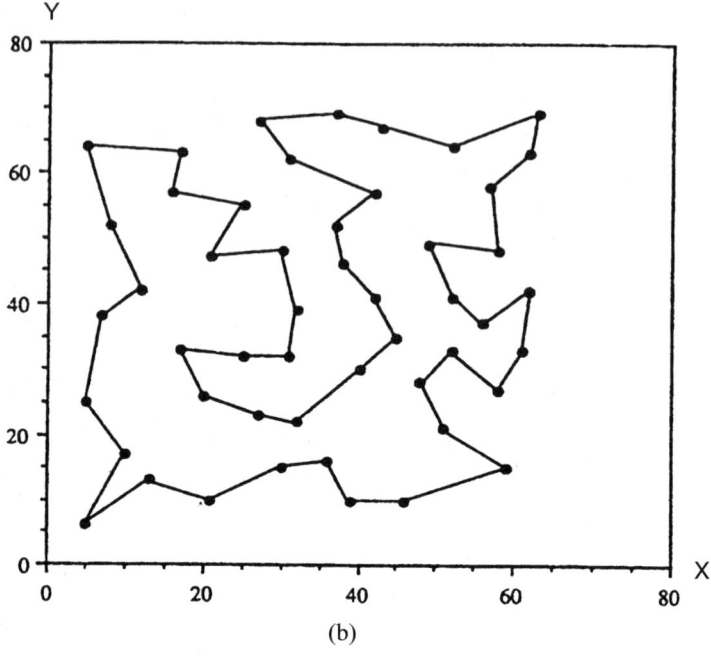

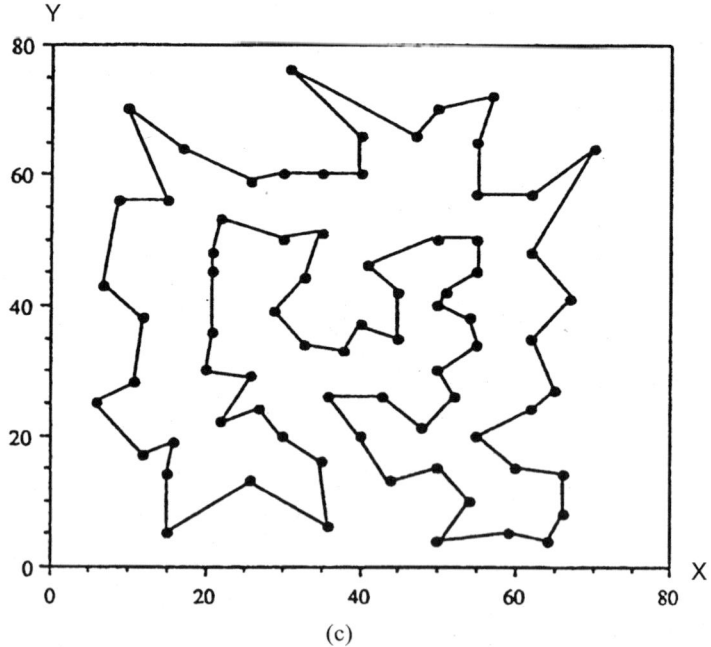

Figure 2.30 (b, c) (continued).

2.9 ROUTING PROBLEMS   79

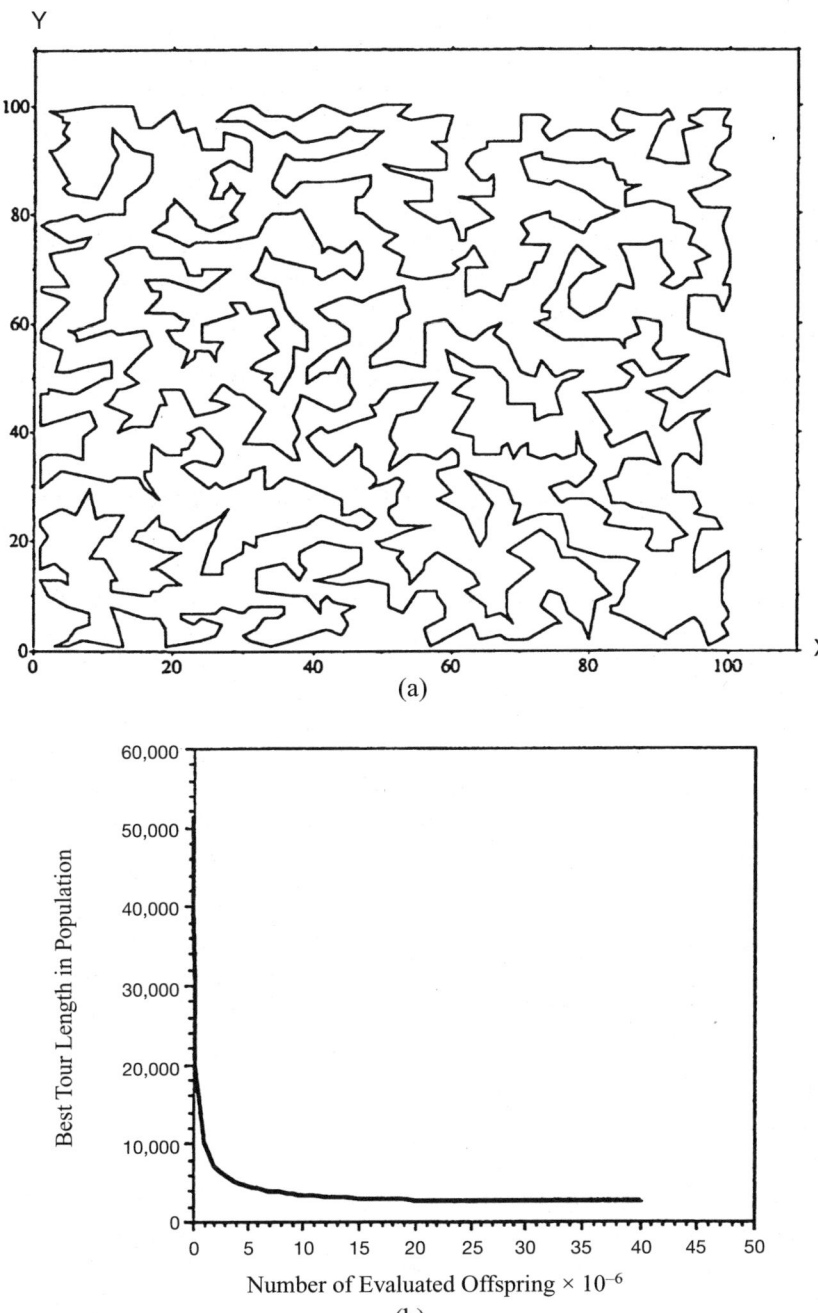

Figure 2.31 (*a*) The best-evolved solution for a 1000-city problem after 40 million evaluated tours. (*b*) The rate of optimization as a function of the number of evaluated solutions. The initial best tour length started at about 52,000 units at the first generation and dropped rapidly (from Fogel, 1993a).

up to 500 cities. A population of 500 parents was used, along with a slightly different form of probabilistic selection. Rather than use the ratio of the two tours' lengths, a win was assigned to the first tour in competition if its length was shorter than its opponent's length. This direct competition was slightly faster computationally, and by adjusting the number of competitors, the overall stringency of selection could be varied from weak (i.e., complete random walk) to strong (i.e., strict deterministic selection). The results indicated a need to examine $O(n^2)$ tours in order to achieve solutions that were on average 10 percent worse than the expected best in uniform TSPs.

Fogel (1990a) also examined the possibility of using evolutionary programming to optimize routes for multiple traveling salesmen (Fig. 2.32), while Fogel and Fogel (1990) evolved solutions to an robotic path problem where single or multiple autonomous vehicles had to be at specified locations at given times, while simultaneously avoiding obstacles (Fig. 2.33).

2.10 COMPARING CROSSOVER, INVERSION, AND MUTATION

Interest in evolutionary computation grew in the late 1980s and early 1990s, particularly with the increased visibility of genetic algorithms (Grefenstette, 1985, 1987; Davis, 1987; Schaffer, 1989; Goldberg, 1989; and others). Several contributions to the literature on genetic algorithms claimed that the key to successful evolutionary optimization was the inclusion of a recombination operator that could exchange building blocks of different individuals (Davis and Steenstrup, 1987; Davis, 1991, pp. 17–18; and others); this followed the view offered in Holland (1975, pp. 110–111) that mutation is essentially an enumerative function and crossover is the primary means for adaptive systems to generate new structures for trial. Some even went so far as to describe evolutionary simulations that relied only on mutation as "shots in the dark" (Bridges and Goldberg, 1987) or examples of "naive evolution" (Schaffer et al., 1989).[7]

Fogel and Atmar (1990) provided an evolutionary programming study to assess the relative importance of mutation, inversion, and crossover when applied to real-valued encodings of solutions to systems of linear equations. Specifically, consideration was given to the system

$$b_i = \sum_{j=1}^{n} a_{ij}(x_j),$$

where the vector **x** represented a coding structure of n "gene" products, the vector **b** represented n phenotypic behavioral responses, and the coefficients a_{ij} represented the respective contribution of each component of **x** to each behavioral response. The quality of a candidate vector **x**, given a matrix A and a desired response vector **b** was judged as the sum of the absolute differences between $A\mathbf{x}$ and **b** term by term.

---

[7]In fairness, Schaffer et al. (1989) acknowledged a stronger role for mutation than had been previously asserted in genetic algorithm literature.

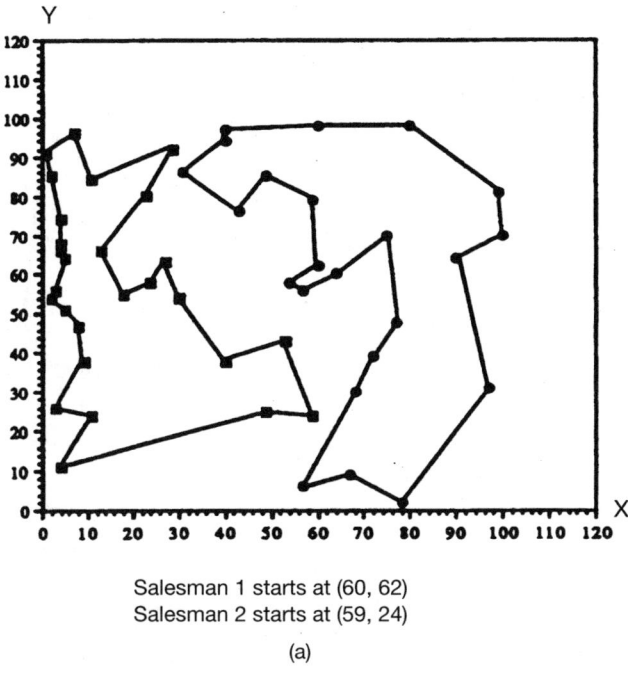

Figure 2.32 (*a*) The best-evolved set of paths for two salesmen with the first starting at point (60, 62) and the second starting at (59, 24). The goal was to minimize the sum of the total path lengths as well as the difference in path length between the two salesmen. (*b*) The rate of optimization as a function of the number of evaluated tours (from Fogel, 1990a).

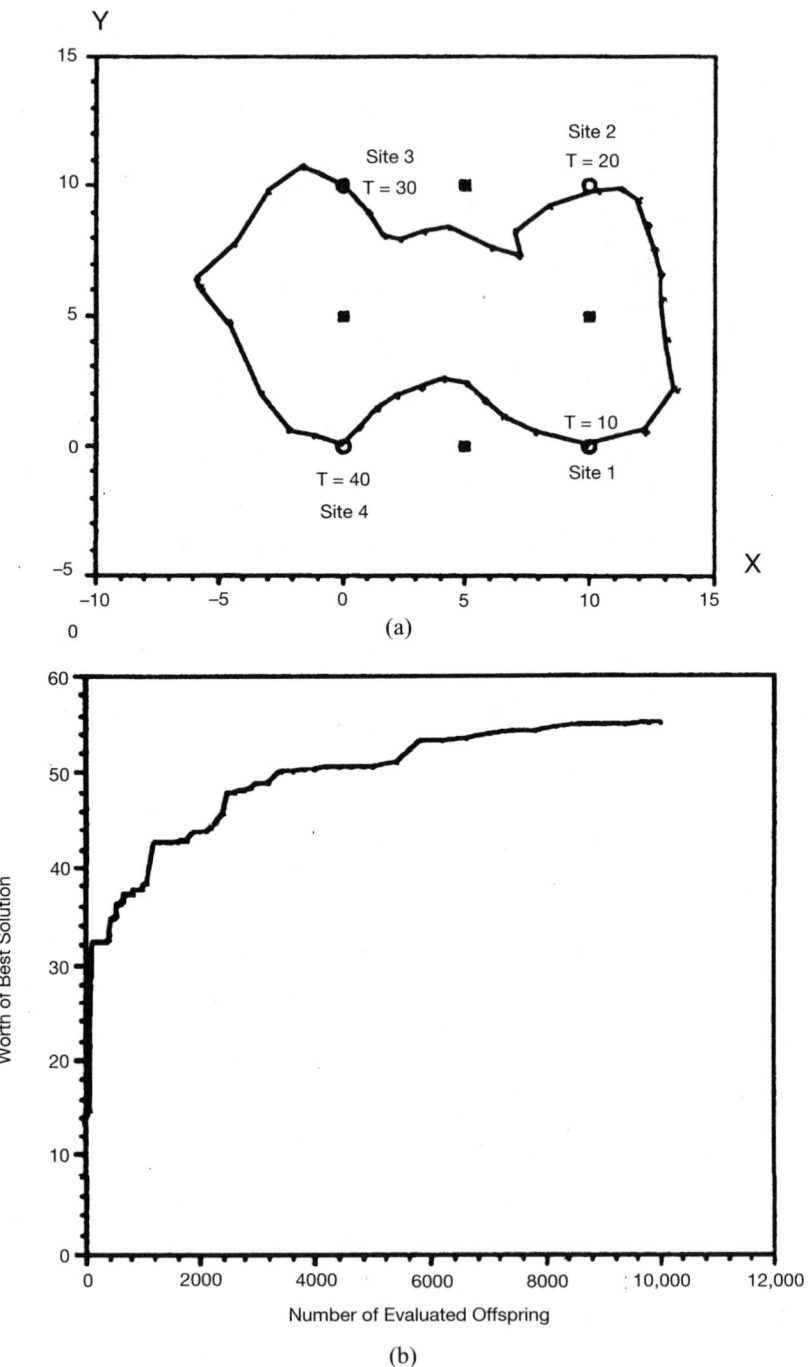

Figure 2.33 (*a*) The best-evolved path for an autonomous vehicle that starts at the origin (0, 0) and must proceed to the four sites in sequence (note that site 4 is at the origin), must arrive at each site at the designated time, and must avoid the locations indicated by squares at (5, 0), (10, 5), (5, 10), and (0, 5). (*b*) The rate of optimization where the quality of solution is judged

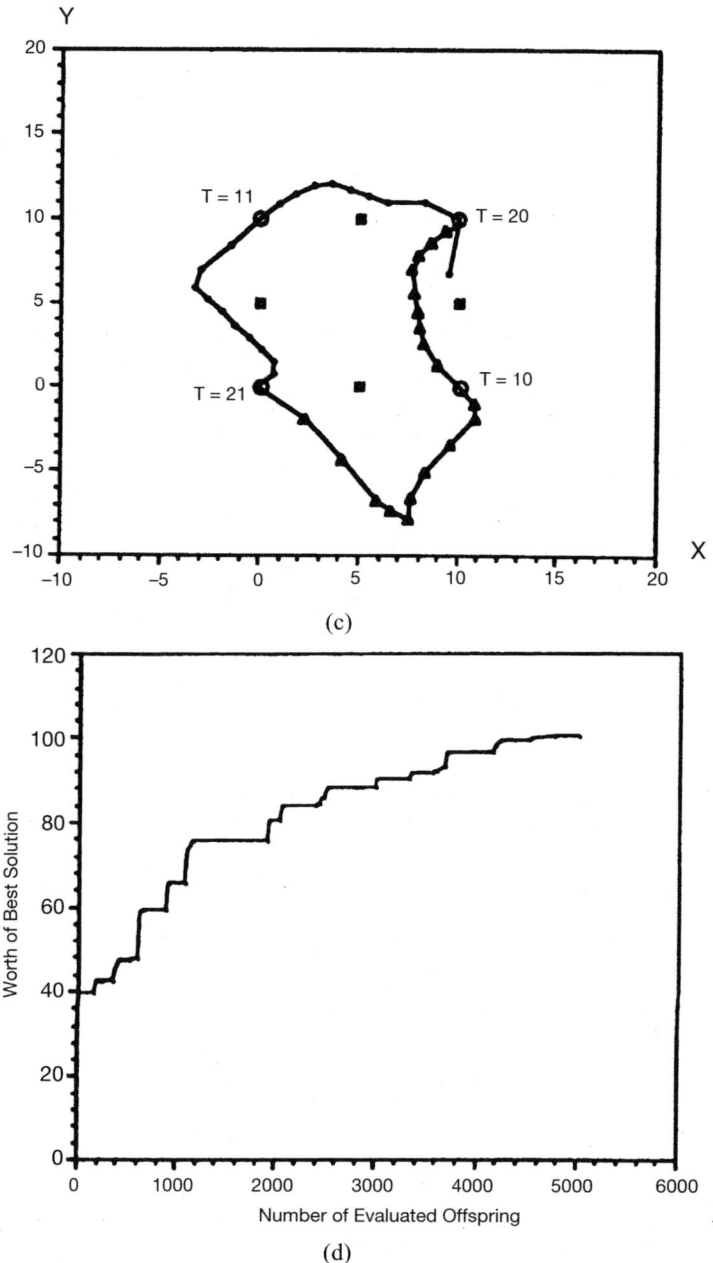

in terms of points received for reaching the targets on time with a penalty for approaching too close to an obstacle. (c) The best evolved paths for two autonomous vehciles where one starts at the origin (0, 0) (the path denoted by the closed circles), and the other starts at (10, 10) (denoted by the triangles). The requirement was again to visit the sites at the specified times while avoiding the obstacles. Note that the first vehicle completes its mission when it passes over site 2 at T = 20 and is not penalized for heading toward the obstacle at (10, 5) after that. (d) The rate of optimization for the multiple vehicle case as a function of the number of evaluated offspring (from Fogel and Fogel, 1990).

The method relied on 150 parent vectors with components initially distributed according to a Gaussian random variable with zero mean and a standard deviation of 30, an arbitrary coefficient matrix $[a_{ij}]$ of rank 10, and an arbitrary desired vector of responses **b**. The 150 initial candidate solutions were assigned to be one of three types (50 each) wherein the first type was subject to variation only by mutation (all coefficients mutated with a standard normal distribution), the second type was varied by one-point crossover (80 percent chance per offspring) and inversion (50 percent chance per offspring), and the third type was varied by crossover, inversion, and also Gaussian mutation with a 1 percent chance per offspring. If genetic operators such as crossover or inversion were to provide an advantage over random mutation alone, it would expected that those vectors that undergo such operations would quickly dominate the population. The results, however, indicated otherwise.

Linear systems of equations provided a convenient method for examining the effects of operators in domains of varying degrees of interactivity. Five sets of trials were conducted where each varied the interactivity of the A matrix by setting the probability of an off-diagonal entry being nonzero to 0, 0.25, 0.5, 0.75, and 1.0, respectively. When the probability is zero, each independent component of a trial vector **x** contributes only to the appropriateness of the corresponding behavioral component. When the probability is 1.0, each component of **x** contributes to the total behavior error summed over all behavioral responses. In each set of trials, 100 repetitions were performed; each was halted after 5000 offspring had been evaluated. The results appear in Table 2.4.

Table 2.4 Percentage of trials where the majority of the population consisted of the given type of vector (defined by the possible variation operators that could be applied)

Degree of Interactivity		Percentage of Trials when 50%+ Consisted of Given Type
0	Random mutation alone	78% (70%)[a]
	Crossover/inversion	9% (5%)
	Crossover/inversion/mutation	13% (4%)
25	Random mutation alone	75% (72%)
	Crossover/inversion	10% (3%)
	Crossover/inversion/mutation	15% (3%)
50	Random mutation alone	83% (83%)
	Crossover/inversion	9% (7%)
	Crossover/inversion/mutation	8% (4%)
75	Random mutation alone	88% (86%)
	Crossover/inversion	2% (0%)
	Crossover/inversion/mutation	10% (6%)
100	Random mutation alone	76% (73%)
	Crossover/inversion	10% (6%)
	Crossover/inversion/mutation	14% (6%)

Source: Fogel and Atmar (1990).

[a]Values in parentheses indicate the percentage of trials in which the given type had completely taken over 100% of the population.

Table 2.5 Using a sign-test on the differences between the best evolved solution to each of the ten trials, no significant evidence existed to suggest that method 2 was superior to method 3. In fact, method 3 discovered a superior solution in six of the ten trials. No solutions discovered by method 1 were superior. The values in parentheses indicate the standard deviations

	Method 1	Method 2	Method 3
	$P(\text{crossover}) = 0.8$	$P(\text{crossover}) = 0.8$	$P(\text{crossover}) = 0.0$
	$P(\text{mutation}) = 0.01$	$P(\text{mutation}) = 1.0$	$P(\text{mutation}) = 1.0$
After 10,000 offspring			
Mean population score	54.19 (±24.404)	21.808 (±3.468)	28.697 (±12.118)
Mean population variance	4.582 (±6.125)	141.725 (±40.372)	170.694 (±203.548)
Mean best score in	52.365 (±22.690)	11.451 (±2.847)	15.176 (±7.084)

Source: Fogel and Atmar (1990).

Advantage quickly accrued to the vectors that were altered by random mutation alone. The observed number of trials in which simple random mutation dominated the population is significantly greater than would be expected under a null hypothesis of the behavior of the evolutionary process being independent of the utilized variation operator ($P < 0.0001$).

Fogel and Atmar (1990) presented a second experiment to investigate the difference in efficiency between systems using crossover and those not using crossover with varying rates of mutation. Again, a system of ten linear equations was used. The matrix A was chosen by setting the entries to random integers between 0 and 9. The desired **b** vector was chosen to make all components of the optimum **x** = 1. Results comparing the effectiveness of different settings for crossover and mutation probabilities are shown in Table 2.5. Again, there was no statistical evidence that any advantage accrued to the use of crossover. Fogel and Atmar (1990) concluded: "While specific circumstances (other than linear equations) may well exist for which crossover and inversion operations are especially appropriate, those conditions cannot be the hallmark of a broadly useful algorithm." This statement has since received both empirical and theoretical justification (Schaffer and Eshelman, 1991; Fogel, 1991b; Bäck and Schwefel, 1993; Rizki et al., 1993; Fogel and Stayton, 1994; Nettleton and Garigliano, 1994; De Jong et al., 1995; Jones, 1995; Eshelman and Schaffer, 1995; Angeline, 1997a; Chellapilla, 1997a; Wolpert and Macready, 1997; Fogel and Ghozeil, 1997; and others).[8] It is now well known that there is no best variation operator to use in all problems or for all representations, and that recombination is simply one of many alternatives that the evolutionary algorithmist may choose from. The appropriateness of the choice depends on the problem.

---

[8]Prior to the conjecture in Holland (1975) on the importance of various operators, Reed et al. (1967) offered a similar observation as to the appropriateness of a particular operator to a particular domain, but it appears that this went unnoticed.

2.11 SUMMARY

The publication of Fogel et al. (1966), and my earlier efforts in evolutionary programming, received significant attention from the artificial intelligence community. The reviews were somewhat mixed (see Feigenbaum, 1963; Kleyn in Fogel et al., 1965c; Solomonoff, 1966; Michie, 1970). But interest from mainstream artificial intelligence in all nonsymbolic methods waned after the publication of Minsky and Papert (1969). Efforts in evolutionary programming were followed up by only a small collection of individuals, yet these efforts diversified from the mid-1960s through the early 1990s into a range of applications and representations.

Rather than force a particular representation onto the problem at hand, attention was given to problem-specific representations and operators tailored to provide efficient searches. The best set of choices for a traveling salesman problem might not be best for solving linear systems of equations or tracking an evading aircraft. Comparisons between different representations and variation operators indicated no advantage for any single choice, a result that was only recently confirmed theoretically (Wolpert and Macready, 1997; Fogel and Ghozeil, 1997). This opened the door to considering a vast array of possible data structures for representing solutions to problems and variation operators that could be applied to transform a parent into an offspring.

The period from 1966 to 1990 included many innovations in evolutionary programming, among them (1) coevolution, (2) probabilistic selection, and (3) adjusting mutation rates as a function of performance. There were also novel attempts to gain a fundamental understanding of evolution via simulations which, in retrospect, might be considered "artificial life." In comparison, mathematical analysis of the convergence properties of evolutionary programming received relatively little attention until the early 1990s. Research in evolving neural networks, fuzzy membership functions, symbolic expressions, and self-adapting these structures awaited future investigation.

CHAPTER 3

SPECIALIZATIONS

3.1 FINDING STRUCTURE IN DATA

One of the key problems in science is the search for patterns in data. The data can be presented in the form of a time series, or as spatial relationships between dimensions (or sometimes both wherein spatial data vary temporally). Some efforts in evolutionary programming in the early 1990s addressed this problem. For example, Fogel (1992b) evolved autoregressive moving-average (ARMA) models of short duration signals, in this case the sound made by ice cracking (also see Fogel, 1990b, 1991b). The problem was of interest to individuals listening for specific signals in the ocean who must differentiate, for example, between human-made and natural sources of sound. The amplitude of the sound wave emanating from an ice crack is shown in Figure 3.1. The objective was to find a parsimonious model of the portion of the time series that corresponds to the crack.

Evolutionary programming was used to search over both the parameters of the possible models as well as the order of the models. An ARMA model can be written as

$$y[t+1] = a_0 y[t] + \ldots + a_k y[t-k] + c_0 e[t] + \ldots + c_l e[t-l], \quad (3.1)$$

where $\{y\}$ is the observed time series, $\{e\}$ is a noise process, $y[t]$ and $e[t]$ represent the observation and noise at time t, and $a_0, \ldots, a_k$ and $c_0, \ldots, c_l$ are coefficients modifying the influence of the corresponding lagged terms in the model. The form is sometimes written as

$$A(q)y(t) = C(q)e(t), \quad (3.2)$$

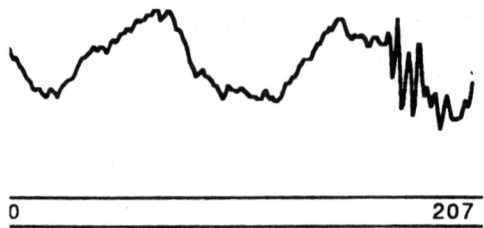

Figure 3.1 Ice crack 1. The event begins near the end of the depicted time series (from Fogel, 1992b).

where all of the AR terms are gathered on the left and the MA terms are gathered on the right, and $A(q)$ and $C(q)$ are polynomials in the shift operator q^{-1}. The problem of determining the proper number of lags k and l and the associated coefficients is typically performed in a two-step process where the investigator first fixes a model order (number of lags) and then employs a gradient search for the best parameters in fitting the available data in terms of minimizing squared error. But this can lead to stagnation at inappropriate parameter settings. In contrast, the evolutionary approach can search over both the coefficients and the number of lags to use simultaneously.

Specifically, three different ice cracks were sampled and served as exemplars for fitting models. The models were represented by a set of parameters in $A(q)$ and $C(q)$, and a number of lags in AR and MA. Each of the initial 250 models was selected with coefficients distributed uniformly over [-0.5, 0.5] and the number of lags in AR and MA to be uniform over the integers [1, ..., 8]. (This choice was somewhat arbitrary.) The cost of each model was chosen to be measured in light of Akaike's information criterion (AIC) (Akaike, 1974). This criterion essentially quantifies the worth of a model by trading off the goodness-of-fit for the number of degrees of freedom (here, the number of lag terms). The criterion is

$$AIC = -2 \ln(f(y|\theta)) + 2p, \tag{3.3}$$

where θ is the parameter vector that defines the model, $f(y|\theta)$ is the likelihood function, and p is the number of independently adjusted parameters. For every parameter that is added, the likelihood function must reflect a significant improvement in the goodness-of-fit; otherwise, a comparison would favor the more parsimonious model even though it possessed greater error. Each model's error was the sum of the AIC scores obtained fitting each of the three ice crack examples.

Offspring were generated (one per parent) by using a zero mean Gaussian mutation applied to every coefficient of each parent, and by adjusting the number of lags in the AR and MA parts of the parent model. Both the AR and MA terms were given a 50 percent chance of being adjusted up or down by a Poisson distributed number of terms, where the Poisson was chosen with rate 0.1. The variance of Gaussian mutations was tuned by hand as an affine function of the parent's error (AIC) score, a process that was later replaced by self-adaptive methods for adjusting the step size of the search on line (see Section 3.2). After all offspring were generated, a tournament was held to determine the best 250 models in the population. These became the new parents, and the process was iterated over 100 generations. The optimization of the error scores for the parent models at each generation is shown in Figure 3.2.

The best-evolved model possessed four AR lags and one MA lag:

$$A(q) = [1 \ -0.7800 \ 0.4040 \ -0.1278 \ -0.3552]$$
$$C(q) = [1 \ 0.5749]$$

and had an overall mean squared error of 0.04020. The goodness-of-fit to the data can be judged visually in Figure 3.3. For comparison, a recursive prediction error method (RPEM) was applied to the same data for the same number of lag terms in $A(q)$ and $C(q)$. This resulted in the model

$$A(q) = [1 \ -1.0113 \ 0.6949 \ -0.4759 \ -0.1056]$$
$$C(q) = [1 \ 0.3525]$$

which had an error of 0.0426. Even though the coefficients for this model are fairly different, interestingly the frequency response for both models are quite similar

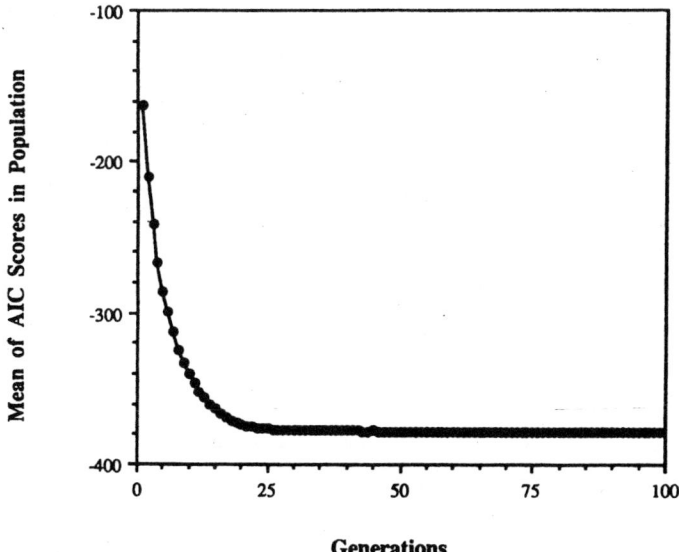

Figure 3.2 Optimization of the mean of the AIC scores in the population as the function of the number of generations while modeling the series of ice cracks. Even though the criterion incorporates both the goodness-of-fit and a penalty term for model complexity, the optimization is seen to proceed quite rapidly (from Fogel, 1992b).

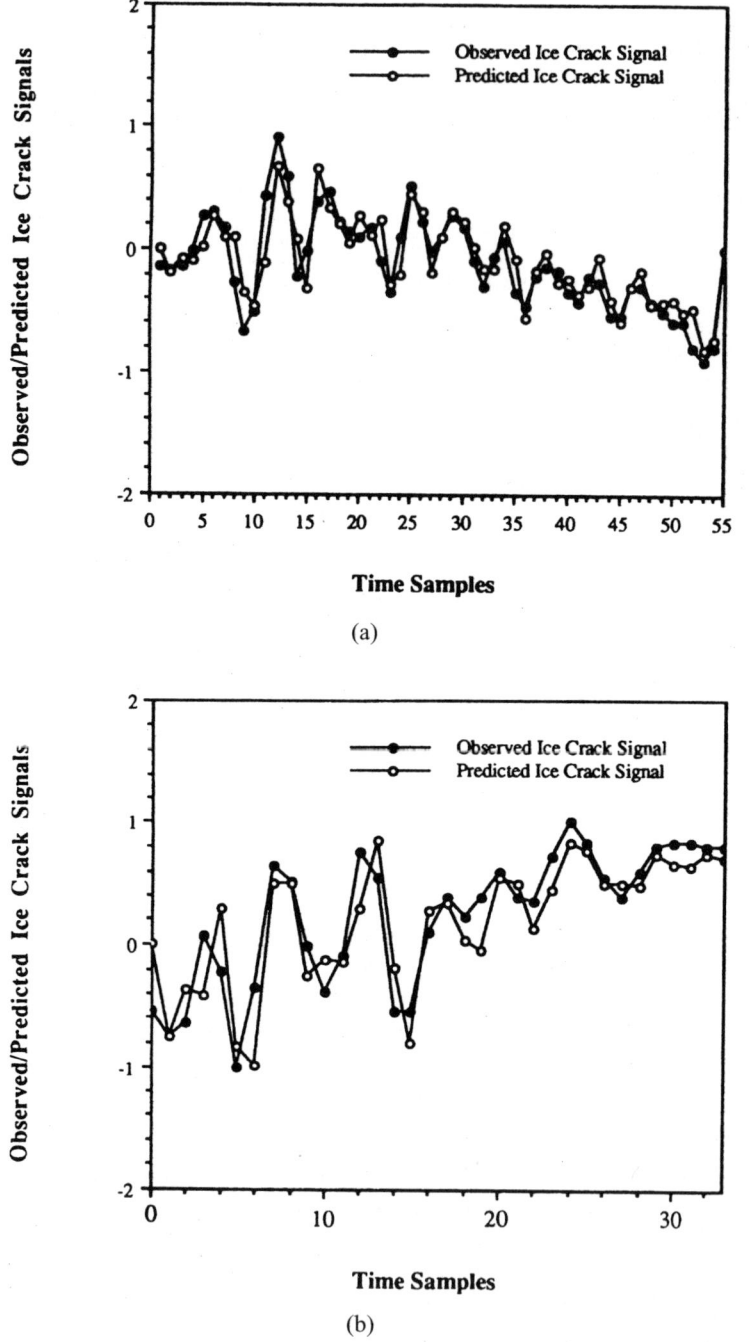

Figure 3.3 Observed and modeled data for (*a*) the first, (*b*) the second, and (*c*) the third ice crack. The model generates a good fit to all three ice cracks (from Fogel, 1992b).

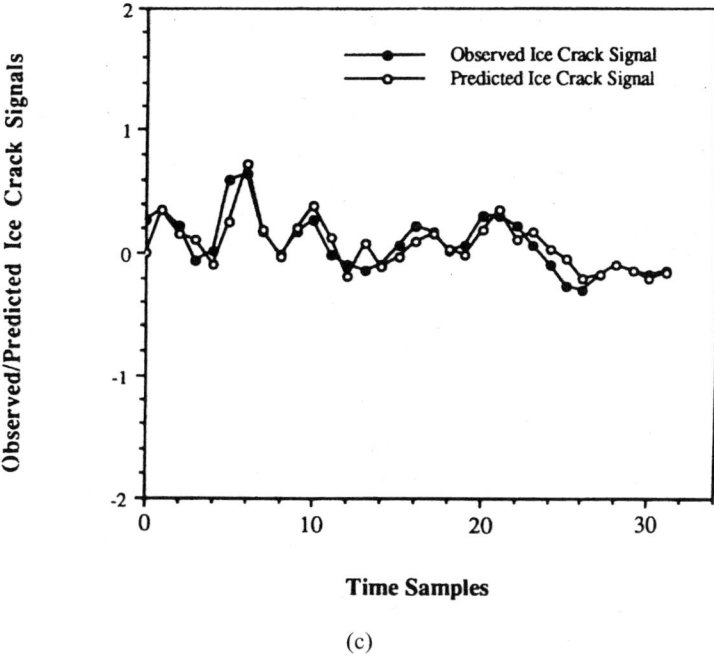

Figure 3.3 (c) (*continued*).

(Fig. 3.4). This points up an important characteristic of modeling systems: The behavior of the coding structure is what is evaluated, not the structure per se. Both the best-evolved model and the model found by RPEM have similar behavior despite having different structures. Attempting to recombine these models would result in failure, which is often the case when the components of complex structures are highly interactive.

Rather than model time series, Fogel and Simpson (1993a, 1993b) examined the problem of finding patterns in spatial data. Earlier, Simpson (1992, 1993) had proposed a procedure for placing hyperboxes around data so as to appropriately cluster and classify patterns of interest. This procedure was deterministic and based on the order of the data as presented. Therefore different clusterings could be obtained if the same data were presented in a different sequence. At first Simpson and Fogel considered evolving the optimal order of presentation of the data (Fogel, 1998a), but this was set aside in favor of an alternative approach that evolved the appropriate size and number of hyperboxes for a given set of data in light of information statistics such as the minimum description length (MDL) principle (Risannen, 1984).

To illustrate, Fogel and Simpson (1993b) evolved hyperboxes to cluster the Fisher Iris data, which has been examined extensively. The data consist of three classes of Iris flower. Each class has 50 exemplars that consist of four separate measurements ranging in value from 0 to just slightly over 9 taken from different portions of

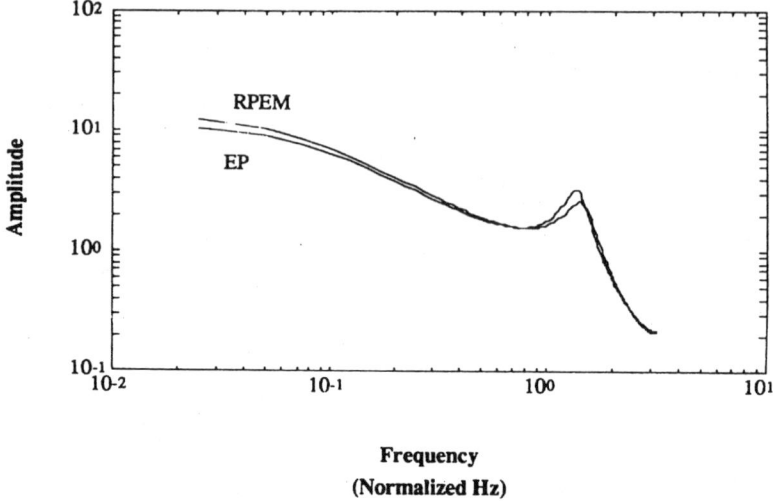

Figure 3.4 The frequency response of both the evolved ARMA model (EP) and the model generated using RPEM. Despite differences in model structure, the frequency response (i.e., the behavior) of both models is notably similar. This emphasizes that function, not form, is the sole measure that undergoes selection in evolution (from Fogel, 1992b).

the Iris flower. The data set consists of 150 four-dimensional patterns. To evaluate the effectiveness of the evolved clusters, each cluster was associated with a class of flower. Ideally the clustering algorithm would find three clusters that uniquely divide the 150 patterns into the original three classes.

The procedure involved a population of 250 solutions, each chosen by first determining a random number of hyperboxes for the selected solution, ranging uniformly between one and five. Two defining points for each hyperbox (min, max) were then selected at random from a uniform distribution ranging between [0, 10] (the natural range of the data). One offspring was created from each parent. The position of all hyperboxes in the selected parent was varied by adding a Gaussian random variable with a mean of zero and standard deviation of 0.1 to every dimension of each defining minimum and maximum point. Further there was a 50 percent chance of varying the number of hyperboxes for each solution. If a change was indicated, there was an equal chance for adding or deleting a hyperbox up to the maximum of five and down to the minimum of one. Hyperboxes to be added were placed randomly (uniform) in the range [0, 10] in each dimension. Hyperboxes to be deleted were selected at random with equal probability.

Parents and offspring were scored with respect to the quality of clustering in an manner analogous to optimal subset selection in regression using the MDL. The MDL criterion is

$$MDL(\theta) = -\log_2(f(\mathbf{x}|\theta)) + 0.5p \log_2(n), \qquad (3.4)$$

where **x** is the observed data, $f(\mathbf{x}|\theta)$ is the conditional likelihood function given a parameter vector θ, p is the number of independently adjusted parameters in θ, and n is the data size. The first term indicates the goodness-of-fit of the clustering, while the second term provides a penalty function for having to use a given number of clusters. The optimum θ minimizes the MDL.

The application of the MDL requires the assumption of a parametric distribution of points in a cluster defined as the interior of a positioned hyperbox. This was assumed to be uniform over the area of the hyperbox. The MDL for any hyperbox was then determined simply by counting the number of data points contained in that hyperbox ($p = 8$ for any hyperbox in four dimensions). After all assigned hyperboxes were assessed, consideration was given to points that were not contained in any hyperbox. These points were treated as outliers drawn from a uniform distribution over the entire available range of the data [0, 10] in each dimension. The total MDL for a given solution was determined by summing each of the individual MDL scores for each hyperbox in that solution, including the hyperbox containing any outliers.

There were a few specific considerations in the use of the MDL that deserve greater attention:

1. Under the assumed uniform distribution, the likelihood for any point in a hyperbox is

$$l = \frac{1}{\Pi_{i=1}^{k}(w_{ji} - v_{ji})}, \quad (3.5)$$

   where there are k dimensions and v_{ji} and w_{ji} are the ith components of the min and max points, respectively, of the jth hyperbox. Thus the MDL will be minimized as negative infinity should any hyperbox converge to bound even a single data point and have zero width in any dimension. The penalty term associated with hyperboxes that contain only one datum ($n = 1$) vanishes. To relieve this problem, the MDL of a hyperbox was only scored when it contained two or more data points. This was intuitively reasonable, since it makes little sense to speak of a "cluster" of only one point.

2. If a hyperbox contained no data ($n = 0$), the penalty term becomes negative infinity and again the MDL score is minimized inappropriately. Requiring a hyperbox to contain at least two points before it will be scored removes this problem, but then extraneous hyperboxes that contain no data incur no penalty. This can be handled in two ways. A small penalty term can be applied to the overall MDL of a solution for each hyperbox regardless of the number of data points each contains, or hyperboxes that contain no data can be pruned at the completion of a clustering run. The latter method was adopted by Fogel and Simpson (1993b).

3. The MDL for any outliers can be assessed even if there is only one such datum. Since the size of the outlier hyperbox is constrained to be the entire available range of the data in each dimension, there is no possibility of driving the MDL score to negative infinity by perfectly fitting a hyperbox to the outlier. The choice of the value of p for the outlier hyperbox may at first ap-

pear problematic because it is constrained to always be the range of the data. Its location is not free to vary. Yet, just like any other hyperbox, it is defined by two parameters in each dimension. Since the MDL is used to assess a fixed clustering, it is reasonable to impose a penalty term for the outlier hyperbox just as for any other hyperbox.

4. The best coding in terms of the MDL depends on the range of the data values. The MDL is composed of two terms: The first assesses the likelihood of the data given a parametric model describing their distribution, while the second imposes a penalty for the degrees of freedom of the model weighted by the length of the data. If the data are scaled to successively smaller values, the penalty term remains the same, but the likelihood function given in the form of (3.4) will grow larger. Therefore the smaller the range of the data, the more hyperboxes will be allowed by the MDL. For the current experiments the data were arbitrarily resized to the range [0, 10].

Probabilistic survival was used based on the typical round-robin competition with 10 competitors, and evolution was halted after 1000 generations. Figure 3.5 shows the mean and best MDL score throughout the evolutionary process. The best-evolved solution used four hyperboxes, but it was discovered that the fourth hyperbox contained no data, and only the remaining three hyperboxes were considered. Figure 3.6 shows the six plots constructed by examining each cluster in every possible pair of dimensions. Some points are not contained in the hyperboxes. However, by using the distance to each cluster as a measure for classification (assigning the

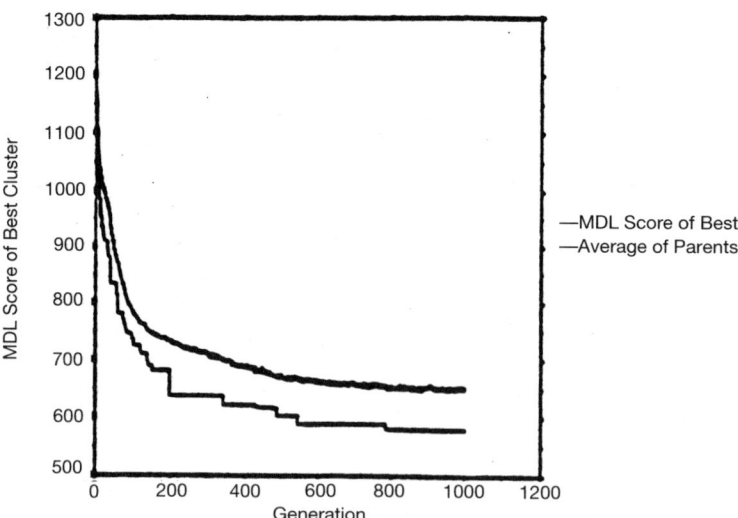

Figure 3.5 The rate of optimization of the best and mean MDL scores in the population as a function of the number of generations when clustering the Iris data with evolving hyperboxes (from Fogel and Simpson, 1993b). The MDL scores start at about 1100 and are reduced to less than 600 within 1000 generations.

3.1 FINDING STRUCTURE IN DATA 95

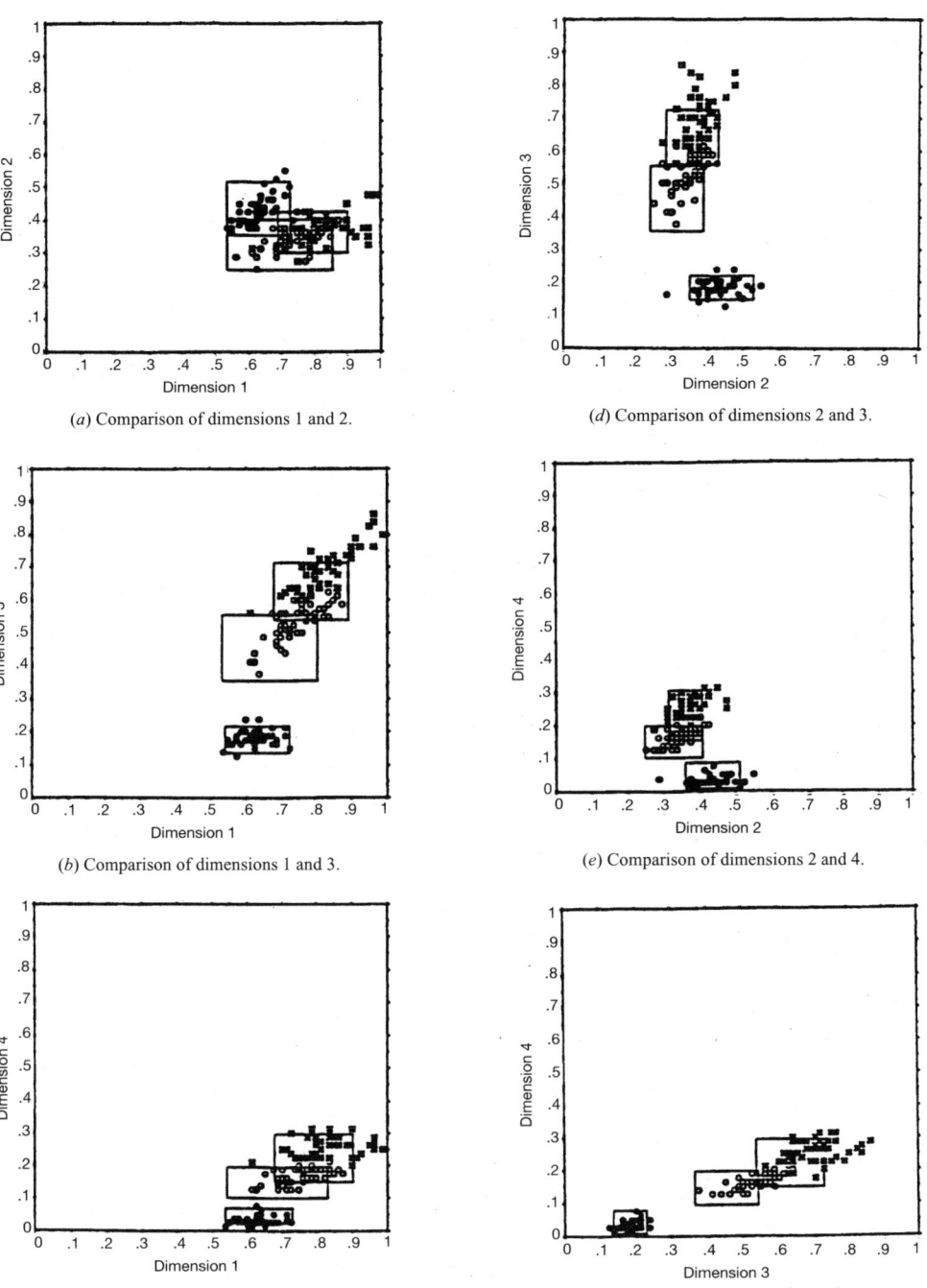

Figure 3.6 The best-evolved clustering of the Iris data from Fogel and Simpson (1993b). The data are scaled to reside within the unit square.

datum to the nearest cluster), the best evolved set of hyperboxes classified all of class 1 correctly and 98% of class 3 correctly. It classified 52% of class 2 correctly and misclassified the remaining 48% as class 3. These results were comparable to the best performance offered by the deterministic method in Simpson (1993) but were not dependent on the order of presentation of the data (i.e., there was no "order" of presentation because the data were considered as a collective whole).

In examining the data in Figure 3.6, it is apparent that they are not separable by boundaries that are aligned with the coordinate axes. Rotatable hyperboxes might be useful in such cases. Ghozeil and Fogel (1996a) implemented this capability and tested it on benchmark cases. Figures 3.7 and 3.8 show the optimization rates and final clusterings for two examples: The first was generated using only one cluster, while the second offered two clusters.

There have been numerous other applications of evolutionary programming to identifying structure in data. Many of these involved time series prediction (Andersen et al., 1991; Rao and Chellapilla, 1996; Fogel and Fogel, 1996; and others), but Sternberg and Reynolds (1997) have used evolutionary programming in combination with cultural algorithms to mine data patterns in automobile insurance claims, looking for patterns that indicate fraudulent applications.

3.2 SELF-ADAPTATION

Given the potential for evolution to optimize solutions to problems, it seems only natural to turn evolution toward the problem of how to best search for these solutions: essentially to use evolution to optimize itself. To make the idea concrete, consider the case where an evolutionary program uses mutation to search for new solutions. The mutation distribution is governed by parameters. Why not change these parameters as information is gained during the evolutionary search (Fogel, 1990)? Why not include these parameters in the evolutionary program so that a search is made for the best solutions and the best mutation procedure for finding those solutions at the same time?

As it turns out, this is an old idea that was implemented in the 1960s (Reed et al., 1967; Rosenberg, 1967; and also apparently within the evolution strategies community) but was unknown in evolutionary programming until the early 1990s. There had been recognition that adapting mutation parameters could be important in accelerating the discovery of suitable solutions (Fogel et al., 1966; Lyle, 1972; and others), but Fogel et al. (1991) offered the first real implementation of a self-adaptive procedure within the framework of evolutionary programming as applied to continuous function optimization.

The task was to find the minimum of a two-dimensional function $f(x, y)$. Two functions were

$$f(x, y) = x^2 + 2y^2 - 0.3 \cos(3\pi) - 0.4 \cos(4\pi) + 0.7 \qquad (3.6)$$

and

$$f(x, y) = 100(x^2 - y)^2 + (1 - x)^2. \qquad (3.7)$$

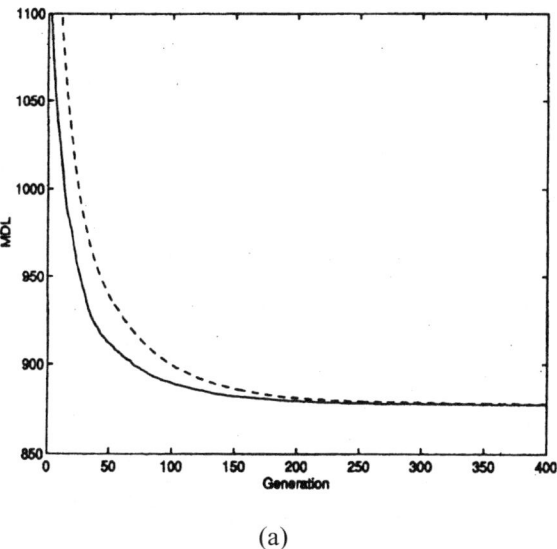

(a)

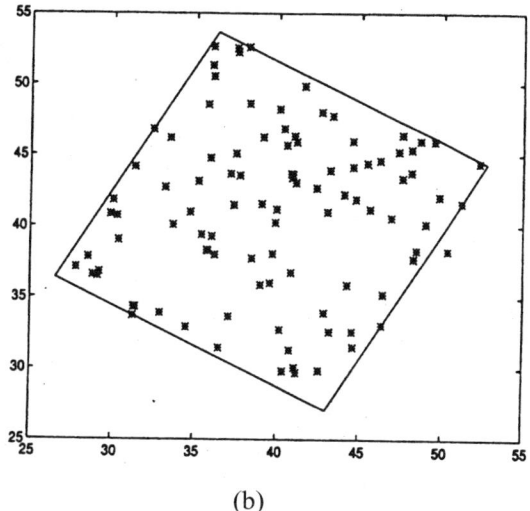

(b)

Figure 3.7 (*a*) The rate of optimization of the best and mean MDL scores in the population as a function of the number of generations when clustering a single rotated box of data. (*b*) The best-evolved solution after 400 generations, which has a single box rotated to fit the observed data. Only one datum remains outside the box. The trade-off for expanding the box to contain this point is too costly. (From Ghozeil and Fogel, 1996.)

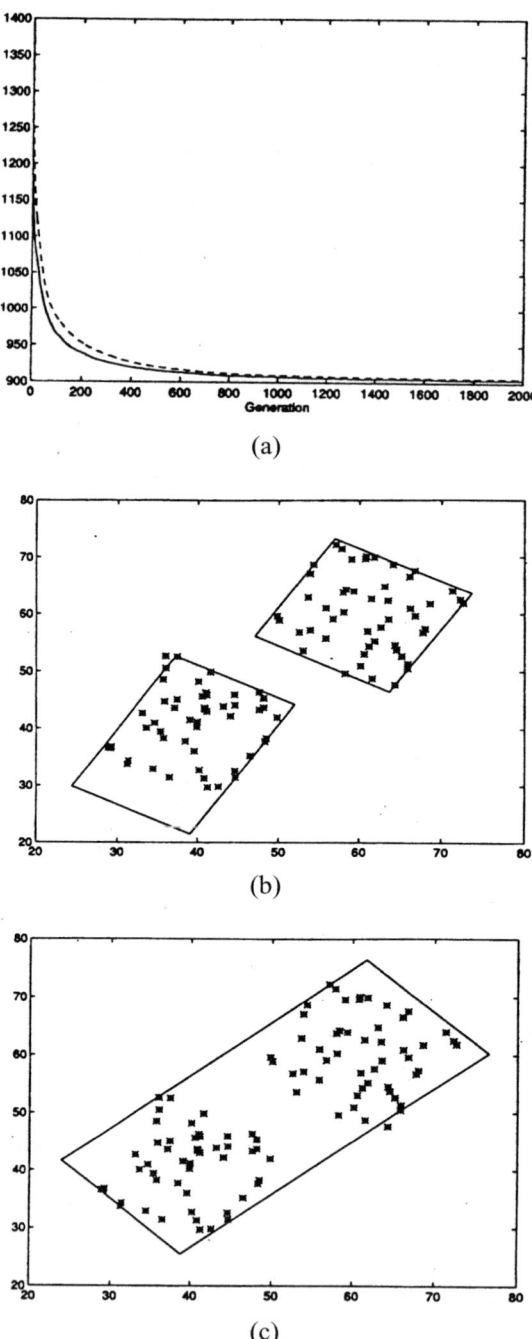

Figure 3.8 (*a*) The rate of optimization of the best and mean MDL scores in the population as a function of the number of generations when clustering a two rotated boxes of data. (*b*) and (*c*) Two best-evolved solutions after 2000 generations. The solution in (*b*) matches the generation of the data by fitting two boxes, but the solution in (*c*) is more parsimonious, using only a single box to cover all the data. The trade-off between parsimony and the large gap between the two clusters generated nearly equivalent scores for these two solutions. (From Ghozeil and Fogel, 1996.)

The first is a multimodal function from Bohachevsky et al. (1986) used to study the effectiveness of simulated annealing, while the second is a unimodal function from Rosenbrock (1960) that has a deep groove in the surface making it difficult for steepest descent methods. Both functions are shown in Figure 3.9. The procedure was implemented as follows:

An initial population of 200 candidate solutions (x_i, y_i), $i = 1, \ldots, 200$ was randomly placed in accordance with a uniform distribution over $[-50, 50]$. Each of these original parents also had a pair of mutative (or perturbation) parameters (px_i, py_i), $i = 1, \ldots, 200$ that were used during the generation of offspring. These parameters were distributed uniformly at random over $[0, 25]$. Each parent was evaluat-

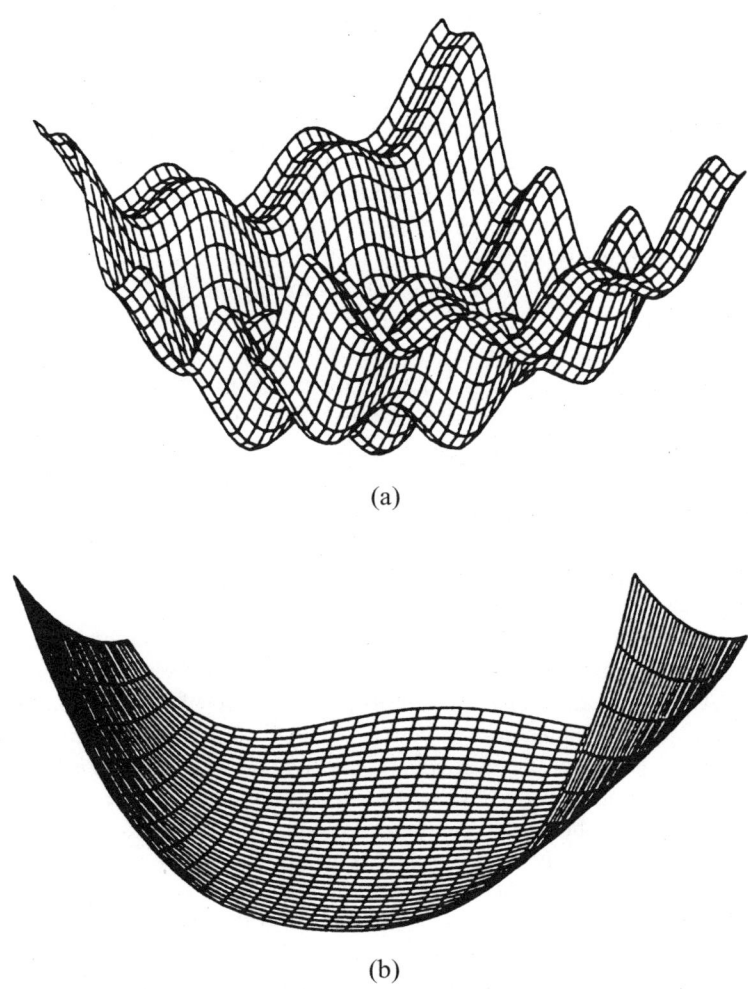

(a)

(b)

Figure 3.9 (*a*) The Bohachevsky function over $[-1, 1]^2$. (*b*) The Rosenbrock function over $[-4, 4]^2$.

ed in light of $f(x_i, y_i)$, with this being taken as their error score. Each parent generated a single offspring (x', y') via Gaussian mutation:

$$x_i' = x_i + N(0, px_i^{1/2}),$$
$$y_i' = y_i + N(0, py_i^{1/2}); \quad (3.8)$$

thus px_i and py_i were the variance of mutation in each dimension. After generating the offspring's Cartesian coordinates, the parent's mutative parameters were copied to the offspring with the addition of Gaussian noise:

$$px_i' = px_i + N\left(0, \frac{px_i}{6}\right),$$
$$py_i' = py_i + N\left(0, \frac{py_i}{6}\right); \quad (3.9)$$

thus the standard deviation for varying the mutative parameters was taken as one-sixth of the parameters themselves. The scaling factor of dividing by six was chosen to ensure that the resulting offspring's mutative parameters would almost always be positive. In the event that either px_i' or py_i' became negative, it was reset to the arbitrary small value of 0.001. Offspring were scored in similar manner as their parents, and tournament selection with ten competitions was applied to select the parents for the next generation (see Chapter 2, Section 2.9). Evolution was halted after 100 generations.

The "meta-evolutionary programming" approach outlined above was executed in 1000 Monte Carlo trials to determine its reliability. To assess any potential advantage to the technique, it was compared with a more traditional evolutionary programming approach where the variance of the Gaussian mutations to Cartesian coordinates was set equal to the parent's error score, along with a binary encoded genetic algorithm with 14 bits of precision in each dimension, one-point crossover applied with probability 0.8, and bit mutation with probability 0.01/bit. Other parameters concerning population size and selection were held constant.

Figures 3.10 and 3.11 show the mean best score in the populations as a function of the number of generations for the Bohachevsky and Rosenbrock functions, respectively. The results showed that the self-adaptive approach to determining the appropriate mutation variance was effective, although the technique was slower to adapt to the Bohachevsky function which was the more traditional approach of setting the variance equal to the parent's error score. Of course the effectiveness of this latter procedure could be rendered nil simply by adding or subtracting a large constant value to the function. In general, it is not possible to know the appropriate heuristic to apply to the parent's error score in generating offspring, and moreover this information is insufficient because different solutions can receive the same score yet need entirely different mutation strategies to generate superior offspring. The self-adaptive approach allows for each parent to evolve specific information on how best to search for successful offspring and requires only a minimal amount of

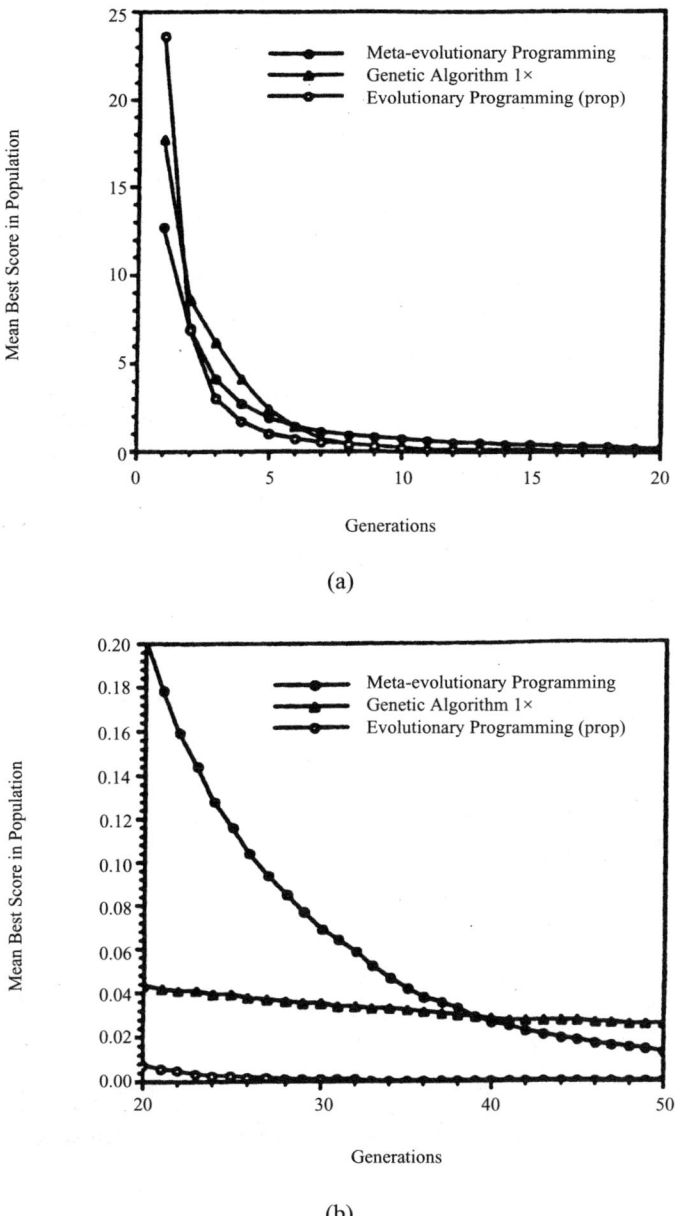

Figure 3.10 (*a–c*) Optimization of the best score in the population averaged over 1000 Monte Carlo trials on the Bohachevsky function with standard evolutionary programming and genetic algorithms, and the self-adaptive evolutionary programming. The three graphs depict different intervals in the course of the evolution (from Fogel et al., 1991). "Meta" refers to the use of self-adaptation, "1×" indicates the use of one-point crossover, and "prop" indicates setting the variance in proportion to the parent's error score.

(*continued*)

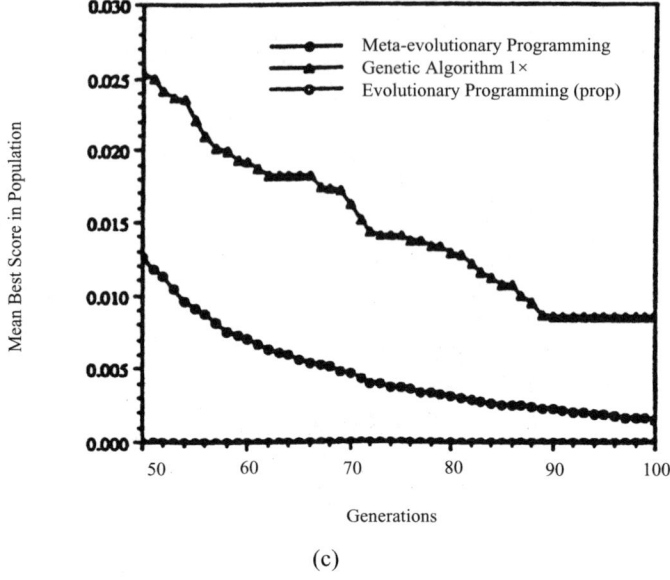

Figure 3.10 (c) (*continued*).

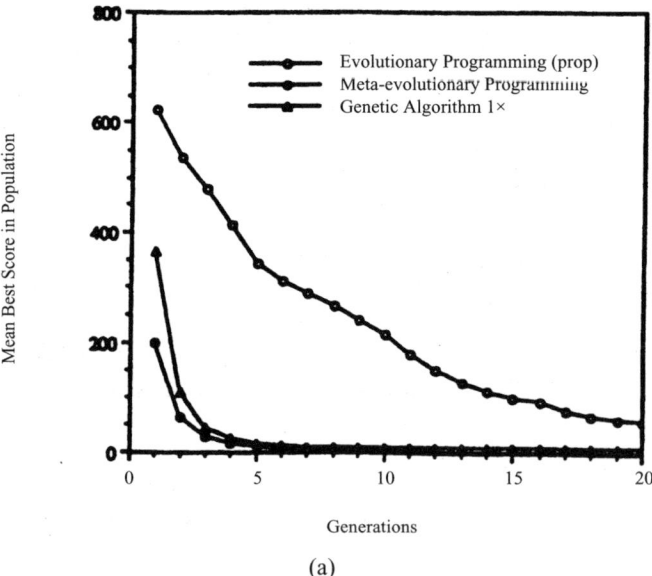

Figure 3.11 (*a–c*) Optimization of the best score in the population averaged over 1000 Monte Carlo trials on the Rosenbrock function with standard evolutionary programming and genetic algorithms, and the self-adaptive evolutionary programming. The three graphs depict different intervals in the course of the evolution (from Fogel et al., 1991). "Meta" refers to the use of self-adaptation, "1×" indicates the use of one-point crossover, and "prop" indicates setting the variance in proportion to the parent's error score.

3.2 SELF-ADAPTATION 103

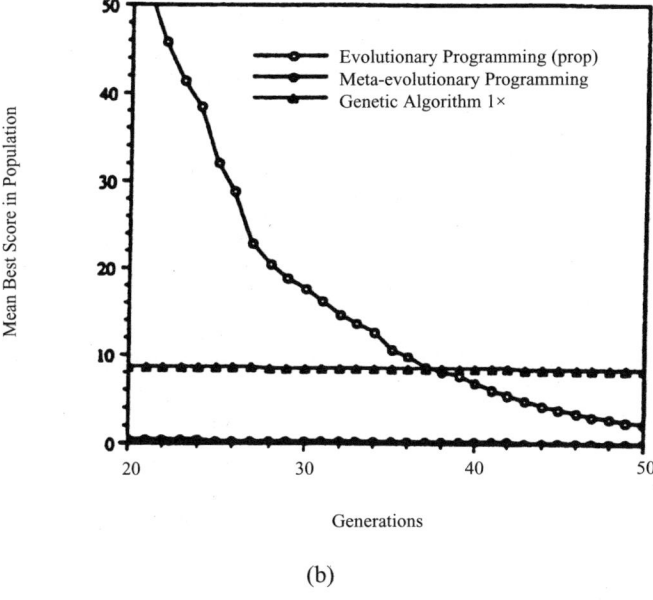

(b)

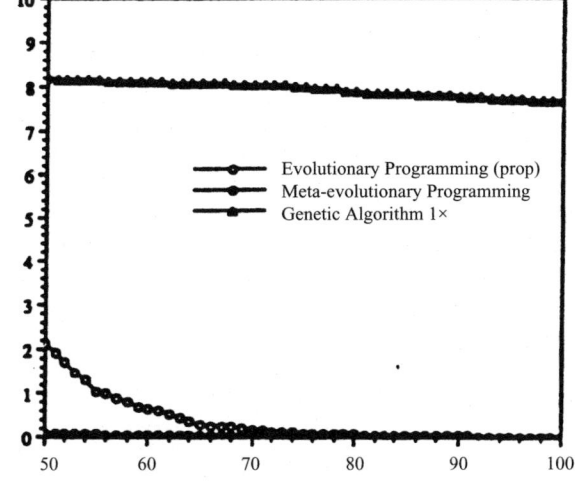

(c)

Figure 3.11 (*b, c*) (*continued*).

human intervention (e.g., in choosing the initialization). Note too that the self-adaptive method found solutions that were statistically significantly better ($P < 10^{-4}$ in both cases) than those found using the genetic algorithm, which prematurely stagnated in both cases.

At the suggestion of Sebald (1991), Fogel et al. (1992) offered a self-adaptive procedure that would allow for correlations between the mutations applied to each dimension. That is, rather than evolving just the variances to apply in Gaussian mutations in x and y, it was possible to also include the covariance of x and y. Thus each individual in the population retained the current (x, y) along with parameters (px, py, r, pr), where r was the correlation between mutations in x and y, and pr was the standard deviation for altering r (in a similar manner as px and py). The values of r and pr were initialized uniformly at random over the ranges $[-1, 1]$ and $[0, 0.1]$, respectively. If the Gaussian mutation of r exceeded either limit of ± 1, the value of r was reset to the limit that it exceeded. This procedure was reasonable for two-dimensional problems but not easily extended to an arbitrary dimensionality (see Saravanan and Fogel, 1996).

The first contact between the German researchers in evolution strategies and the evolutionary programming community was made shortly before the First Annual Conference on Evolutionary Programming in February 1992 (Bäck, 1992). The similarities between the two approaches when treating continuous optimization problems were clearly evident, including the inclusion of self-adaptation. There was a slight difference in that the procedure adopted within evolution strategies relied on lognormal perturbations of the standard deviations governing the step sizes of Gaussian mutation. That is, under evolution strategies, the self-adaptive parameters were updated as

$$\sigma_i' = \sigma_i \times \exp(\tau N(0, 1) + (\tau' N_i(0, 1)), \tag{3.10}$$

where i denotes the dimension, σ_i is the standard deviation for the ith dimension, $N(0, 1)$ is a standard Gaussian random variable sampled once for all i dimensions, $N_i(0, 1)$ is a standard Gaussian sampled anew for each dimension i, and τ and τ' are scaling terms. By comparison, using the same notation, under evolutionary programming the procedure was

$$\sigma_i' = \sigma_i + \frac{\sigma_i'}{6} N(0, 1). \tag{3.11}$$

Another slight difference was that under evolution strategies, offspring coordinates were generated using the mutated step sizes

$$x_i' = x_i + \sigma_i' N(0, 1), \tag{3.12}$$

where x_i is the ith coordinate of the parent, whereas in evolutionary programming, offspring coordinates were generated before updating the mutation step sizes

$$x_i' = x_i + \sigma_i N(0, 1). \tag{3.13}$$

It was natural to compare these two methods of self-adaptation on different function optimization problems. The first such comparison was offered in Bäck and Schwefel (1993), which appeared to favor the lognormal perturbation method, but unfortunately the scaling term of dividing by six for the Gaussian method was incorrectly implemented as multiplying by six. This would have had the effect of quickly making the variances become negative, thereby being reset to only a very small perturbation.[1] Saravanan and Fogel (1994) and Saravanan et al. (1995) offered more legitimate comparisons, and these again generally favored the lognormal method of updating the self-adaptive parameters (only in one of six cases studied did the Gaussian method statistically significantly outperform the lognormal procedure). As a consequence the lognormal procedure was adopted generally within evolutionary programming (Gehlhaar and Fogel, 1996; Fogel et al. 1997; and many others). Interestingly Angeline (1996) showed that the Gaussian method can outperform the lognormal procedure when sufficient noise is imposed on the response surface being searched. Thus the question of when either method might be more appropriate, and for which functions, still remains open. And there is also an open question of whether there might be some benefit from self-adapting directional mutations in polar coordinates (r, θ) rather than Cartesian coordinates (x, y) (Ghozeil and Fogel, 1996b).

The use of self-adaptation in evolutionary programming is now routine, but it is not restricted to continuous optimization problems. Discrete combinatorial problems can also be addressed using the technique. For example, Fogel et al. (1994, 1995) and Angeline et al. (1996) used self-adaptation to adjust the mutation probabilities of finite state machines on-the-fly as they predicted a sequence of symbols (as in Fogel et al., 1966). This was done both by (1) adjusting overall probabilities of adding or deleting states, or changing output symbols, state-transitions, or the start state, and (2) incorporating parameters for every state, transition, and output symbol for every machine. The former offered an "individual level" adaptation, whereas the latter offered "component level" adaptation. Recall that the previous mutation strategy was to assign equal probabilities across all modes of mutation, and then again equal probabilities for each component (e.g., if the selected mutation was to change an output symbol for some input in, say, state 1, each possible input symbol would have equal probably of being chosen). A variety of mechanisms of self-adaptation were explored and compared to the use of a fixed probability of various mutations. The results favored the use of self-adaptation (see Figs. 3.12 and 3.13).

In another example, Chellapilla and Fogel (1997a) used self-adaptation to guide the search for superior solutions to traveling salesman problems (TSPs). Previous efforts in Fogel (1993b; and others) relied on a representation comprising an ordered list of the n cities. Variation was applied by choosing two cities along the list at random and reversing the segment of cities between these two points. The effec-

---

[1] It is interesting that even with this problematic implementation, the self-adaptive evolutionary programming was seen to outperform the genetic algorithm on many of the test functions examined in Bäck and Schwefel (1993).

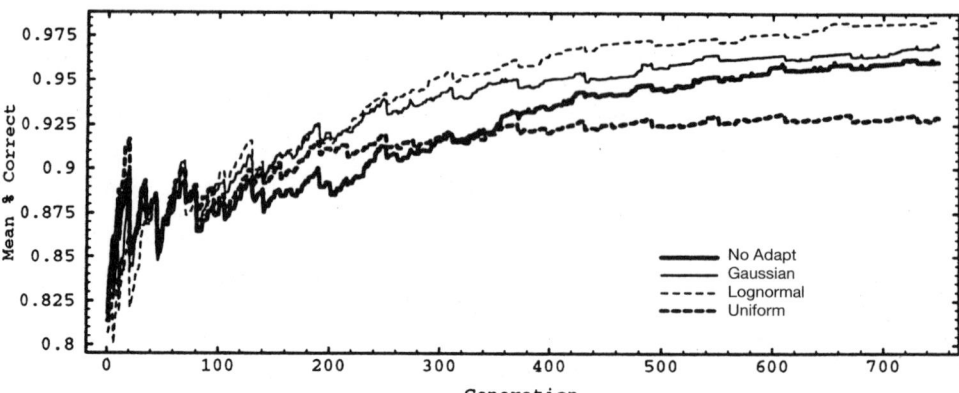

Figure 3.12 Results from Angeline et al. (1996) comparing the cumulative fraction of correct predictions of the sequence (101110011101)* made by an evolutionary program when the mutation distribution was controlled by various mechanisms. The lognormal and Gaussian self-adaptation of the mutations to finite state machines appeared to generate the best results. The "uniform" method simply randomized the mutation probabilities at random for each offspring, thereby limiting the inheritance from the parent. The poor performance of this approach might be expected. The curves are the means taken over 50 independent trials with each method.

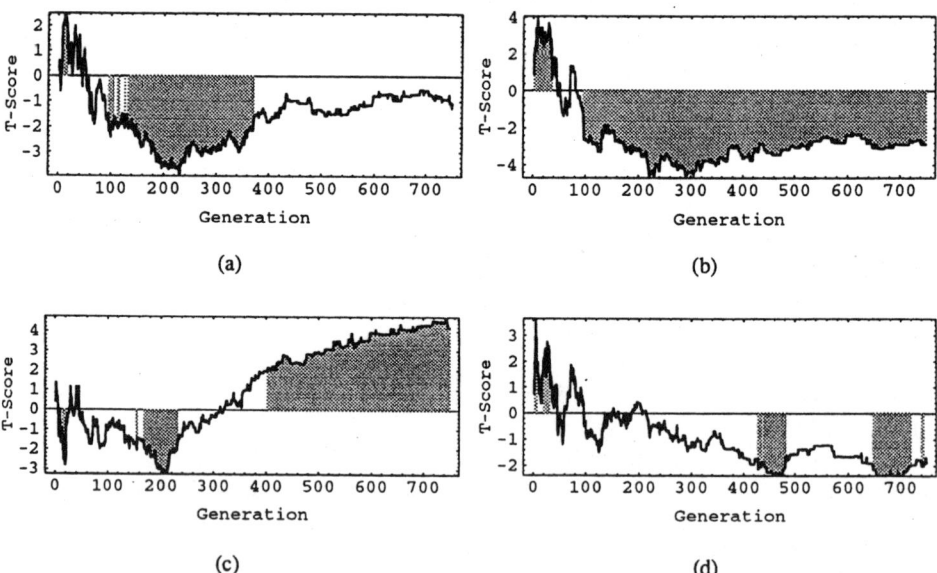

Figure 3.13 T-tests for four pairings in the experiment shown in Figure 3.12. Grey regions denote t-score that generated a P-value less than 0.05. But note that the data at each generation are not independent, so no statistical significance is claimed here. A positive value in the graph denotes a higher mean for the first method, and a negative value shows an advantage for the second. (a) No adaptation vs. Gaussian. (b) No adaptation vs. lognormal. (c) No adaptation vs. uniform. (d) Gaussian vs. lognormal. (From Angeline et al., 1996.)

tive change to the tour depends in part on the length of such a reversal; thus it is natural to imagine that this length could be tuned to the problem as evolution proceeds. Moreover the degree of severity of the mutation could be increased if multiple inversions were imposed on a parent when generating an offspring. Chellapilla and Fogel (1997a) implemented two versions of self-adaptation where in the first, the length of a single inversion was self-adaptive, while in the second, all possible inversions of length 1 to $n/2 - 1$ (due to the problem's symmetry, larger inversion need not be treated) were assigned independent self-adaptive probabilities of occurring. The average results over 30 trials with 100-city TSPs indicated an advantage for the former method (Fig. 3.14).

Returning to continuous optimization problems, there has been recent interest in using Cauchy mutations in place of Gaussian mutations because they offer consistently longer jumps (Fig. 3.15) and may therefore be better suited for escaping local optima (Yao and Liu, 1996). On the other hand, Cauchy mutations are less suited for local optimization (Yao et al., 1997). Rather than choose one or the other a priori, or generate offspring from both simultaneously and then choose the better performing offspring (Yao et al., 1997; Yao and Liu, 1998a), another possibility is to self-adapt the likelihoods of using either operator, or blend the two operators.

Saravanan and Fogel (1997) investigated the former idea of using self-adaptation to probabilistically choose, on-the-fly, whether a parent would generate an offspring by Gaussian or Cauchy mutation. Algorithmically each solution in the

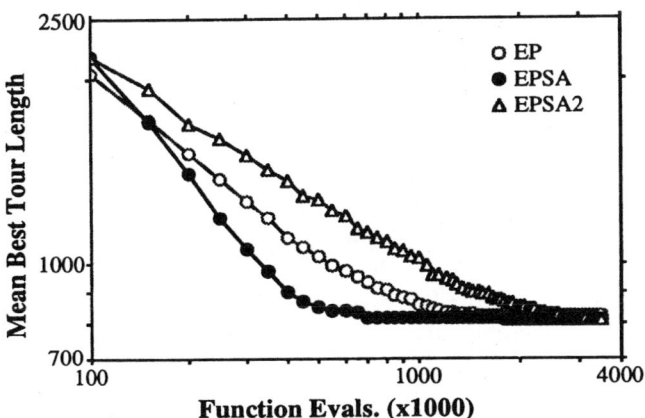

Figure 3.14 The results from Chellapilla and Fogel (1997a) comparing non-self-adaptive evolutionary programming on 100-city traveling salesman problems to two methods for self-adapting the variation operator (described in the text). The single-step self-adaptive method (EPSA) provided the most rapid initial performance, while the multistep method offered the least rate of optimization. The multistep method eventually outperformed both the single step and the standard non-self-adaptive method.

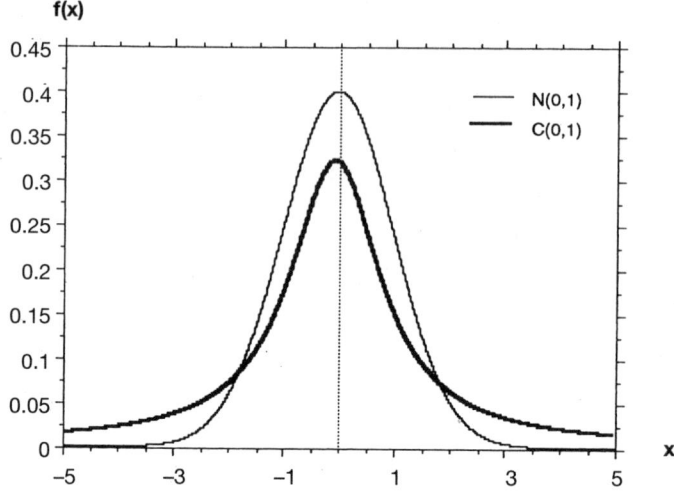

Figure 3.15 The probability density functions for standard Gaussian and Cauchy distributions. The fatter tails of the Cauchy distribution result in more samples that are farther away from the center of the distribution than occur under the Gaussian distribution.

population was described by a 5-tuple ($\mathbf{x}_i$, $\sigma\mathbf{c}_i$, $\sigma\mathbf{g}_i$, $\mathbf{p}_i$, $\mathbf{p}\sigma_i$), $i = 1, \ldots, \mu$, where there were μ parents. The vector $\mathbf{x}_i$ (of n parameters) was the vector of "objective" variables (the variables that are placed in the objective function and are scored), vectors $\sigma\mathbf{c}_i$ and $\sigma\mathbf{g}_i$ were the step sizes used to update the objective variables (for the Cauchy and Gaussian operators, respectively), $\mathbf{p}_i$ was a real-valued vector of length two, one component for each operator, that was used to calculate the probability of applying either the Cauchy or the Gaussian operator, and $\mathbf{p}\sigma_i$ was another vector of length two defining the step sizes that were used to update the vector $\mathbf{p}_i$.

Offspring were generated as follows: For $j = 1, 2$ (one for each operator),

$$p\sigma'_{ij} = p\sigma_{ij} + \alpha N_j(0, 1),$$
$$p'_{ij} = p_{ij} + p\sigma'_{ij} N_j(0, 1), \qquad (3.14)$$

where α was chosen as $(2)^{-0.5}$. If either $p\sigma'_{ij}$ or p'_{ij} were negative, they were reset to a small value (0.0001). The probability of using the Gaussian mutation was

$$P_G = \frac{p'_{i1}}{p'_{i1} + p'_{i2}}, \qquad (3.15)$$

with the probability of using the Cauchy mutation then being the complement

$$P_c = 1 - P_G. \qquad (3.16)$$

In the case where the Gaussian operator was chosen, the step sizes for that operator were updated using the lognormal self-adaptation approach,

$$\sigma g'_{ij} = \sigma g_{ij} \times \exp(\tau N(0, 1) + \tau' N_i(0, 1)), \tag{3.17}$$

while the step sizes for the Cauchy operator were decayed (i.e., reduced by a factor),

$$\sigma c'_{ij} = \beta \sigma c_{ij} \tag{3.18}$$

with $\beta = 0.95$. The objective variables were then mutated using the updated step size vector:

$$x'_{ij} = x_{ij} + \sigma g'_{ij} N_j(0, 1) \tag{3.19}$$

Similarly, if the Cauchy distribution were chosen as the mutation operator, the objective variables would be updated using step sizes σc_i, and the Gaussian step sizes would be decayed. The reason for decaying the unused mutation operator's step size was that this increased the probability that its next use would be successful (smaller step sizes are more likely to result in improvements in many studied test functions).

This procedure was compared with the use of just a Cauchy mutation operator on three test functions, and the results showed that allowing the evolutionary program to choose between the Cauchy and Gaussian mutations offered an advantage (Fig. 3.16).

Alternatively, Chellapilla and Fogel (1997b) offered a procedure where offspring were created from the mean of a Cauchy and Gaussian mutation of the parent:

$$x'_{ij} = x_{ij} + \tfrac{1}{2}\sigma'_{ij}(C_j(0, 1) + N_j(0, 1)), \tag{3.20}$$

essentially using the convolution of these two operators to define a new mutation distributions. Also studied was an adaptive version of this mutation:

$$x'_{ij} = x_{ij} + \alpha' C_j(0, 1) + \beta' N_j(0, 1), \tag{3.21}$$

where the contribution of the Cauchy and Gaussian operators can be scaled. This latter version can then generate exclusively Cauchy or Gaussian mutations, or any linear mixture of these two distributions.

Another interesting variation of self-adaptation was explored in Davis (1994), where histograms were evolved that described the probabilities of mutating a given parameter by a specified amount. Each parent was accompanied by such a histogram, each of 100 bars, which was effectively a probability mass function over all possible mutation steps to be applied in the generation of an offspring. The histogram bars were themselves subject to mutation and therefore could evolve arbitrary shapes. In some cases the ultimate form of these histograms appeared to ap-

110 SPECIALIZATIONS

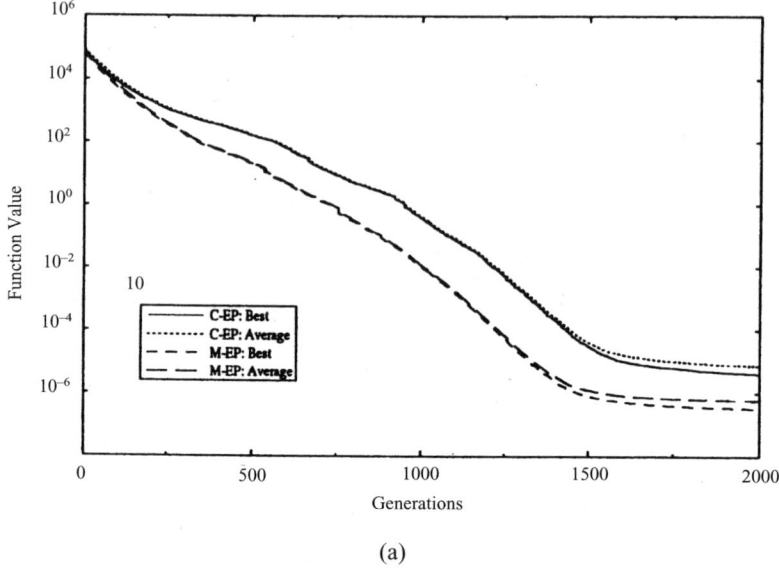

(a)

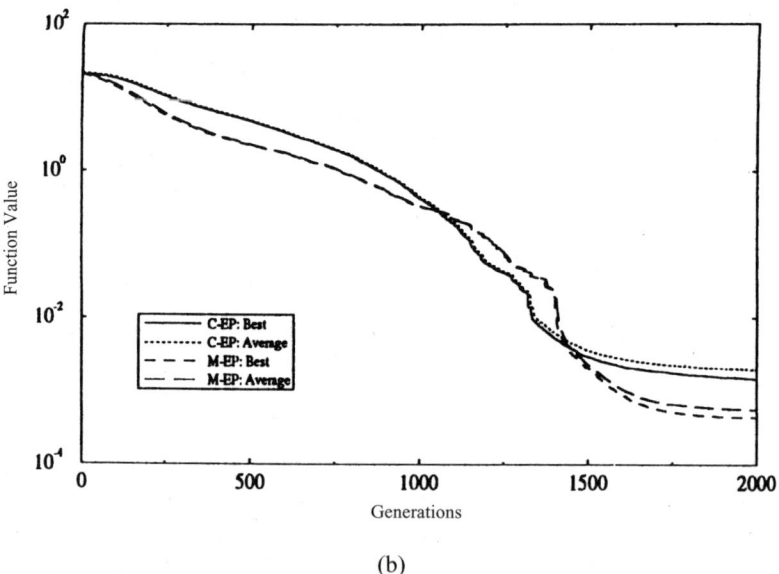

(b)

Figure 3.16 The mean results over 50 independent trials from Saravanan and Fogel (1997) comparing evolutionary programming with Cauchy mutations alone (C-EP) to the self-adaptive method of choosing between Cauchy and Gaussian mutations (M-EP) described in the text. (a) Results for the quadratic bowl, (b) results for Ackley's function, and (c) results for the Rosenbrock function. All functions were examined in 30 dimensions. The results tended to favor the self-adaptive approach ($P < 0.0001$).

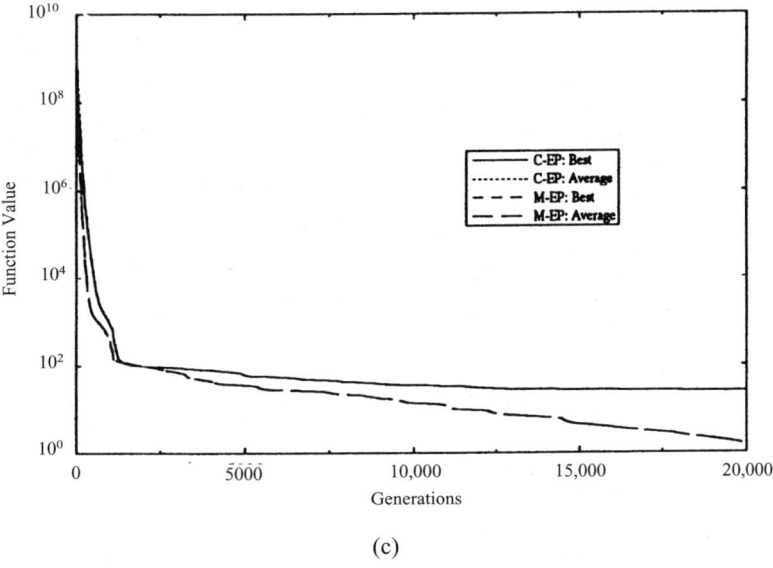

(c)

Figure 3.16 (*continued*).

proach a Gaussian distribution (Fig. 3.17). This distribution-free self-adaptation has received essentially no further attention in the literature, yet it appears to be a promising method toward allowing an evolutionary optimization to be completely free to adapt to the response surface at hand without a priori knowledge.

Since first being introduced in evolutionary programming in 1991, self-adaptation has become a routine element in the general algorithm. For other examples, see Chung and Reynolds (1997) and Porto et al. (1998).

3.3 EVOLVING NEURAL NETWORKS

A natural outgrowth of the extension of evolutionary programming from treating finite state machines to handling real-valued vectors was its application to the optimization of connection strengths in neural networks. Some of the first efforts along these lines in evolutionary programming were Fogel et al. (1990a, 1990b). These initial investigations were limited to single hidden layer feedforward perceptrons. Simple pattern recognition tasks were selected such as the exclusive-or function and discrimination between samples from a Gaussian distribution with mean zero and another with mean 1.0. In essence the task was similar to the optimization of polynomial coefficients undertaken earlier in Fogel and Fogel (1986), in that the hidden nodes in the neural networks under consideration could be written as functions of the form

112 SPECIALIZATIONS

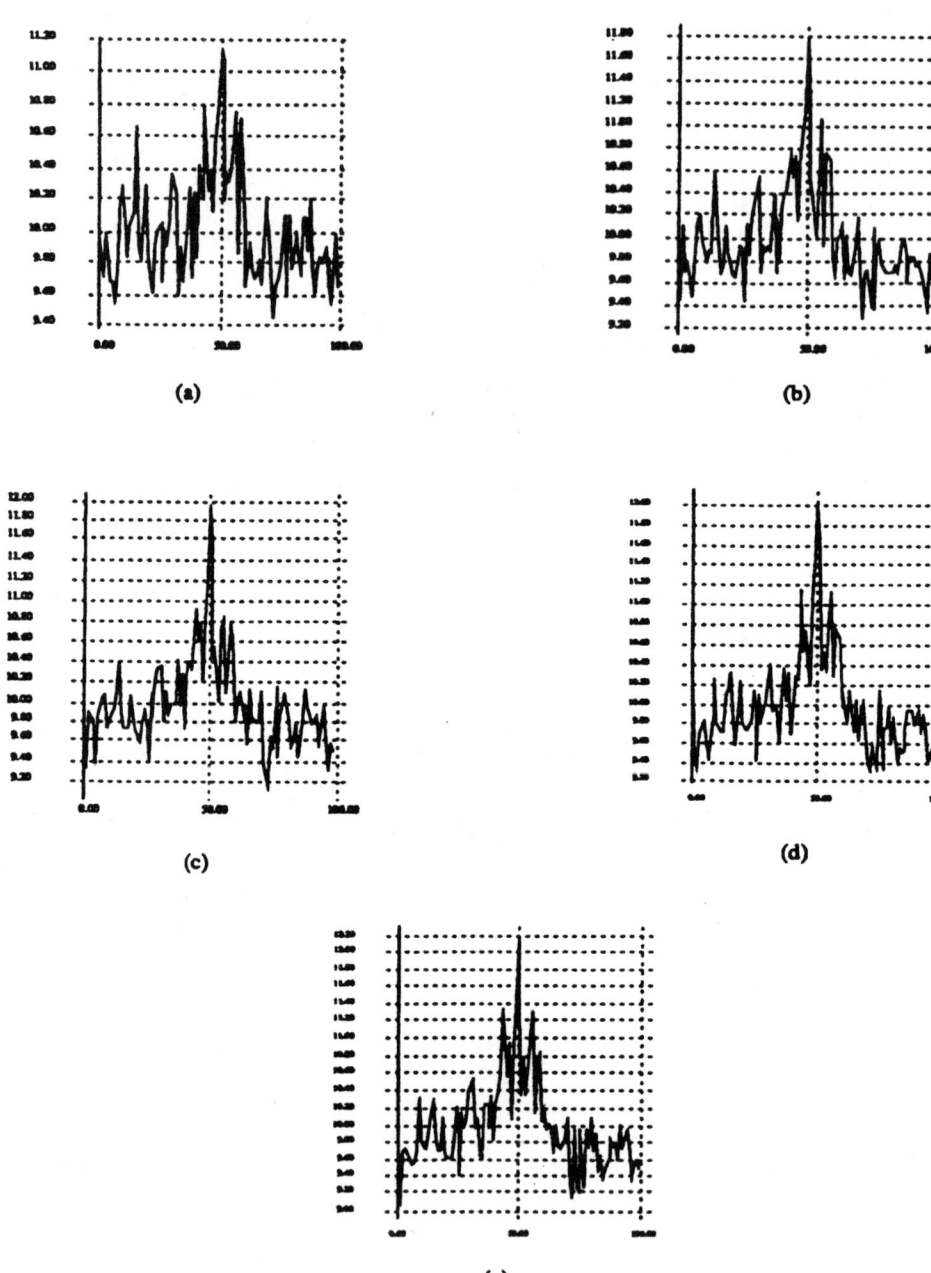

Figure 3.17 Mutation histograms of the best solutions averaged over 100 trials at generations (*a*) 200, (*b*) 300, (*c*) 400, (*d*) 500, and (*e*) 600 on a function with five minima in two dimensions (an expanded Bohachevsky function) from Davis (1994). The distributions appear to be approximately Gaussian distributed.

$$b_j = f\left(\sum_{i=1}^{n} w_{ij} a_i - \theta_j\right), \tag{3.22}$$

where b_j is the output of the jth hidden node, w_{ji} is the weight connecting nodes i and j, a_i is the activation from node i, θ_j is the threshold value for node j, and f is a nonlinear function, typically sigmoid so that

$$f(z) = \frac{1}{1 + \exp(-z)}. \tag{3.23}$$

The final output of the network was calculated as the weighted sum of the values b_j for all hidden nodes j. Thus any matrix of values w_{ij} for a neural network (typically written in vector form) could be evaluated in light of the appropriateness of the input-output behavior of the network given a set of input patterns and desired output values.

Each network's weights were encoded as a vector of real values, and zero mean Gaussian mutation was applied to each weight to create offspring from each parent. The variance of the mutations was often set proportional to the squared error of the parent network, but this was later supplanted with the self-adaptive procedure outlined in the previous section.

The first real-world application of this technique was offered in Porto (1990) in which neural networks were evolved to classify active sonar returns from metal objects (emulating mines) as being different from background reverberations. A typical raw continuous transmission frequency modulation (CTFM) active sonar return is shown in Figure 3.18. Following Porto (1989), attention was devoted to four features

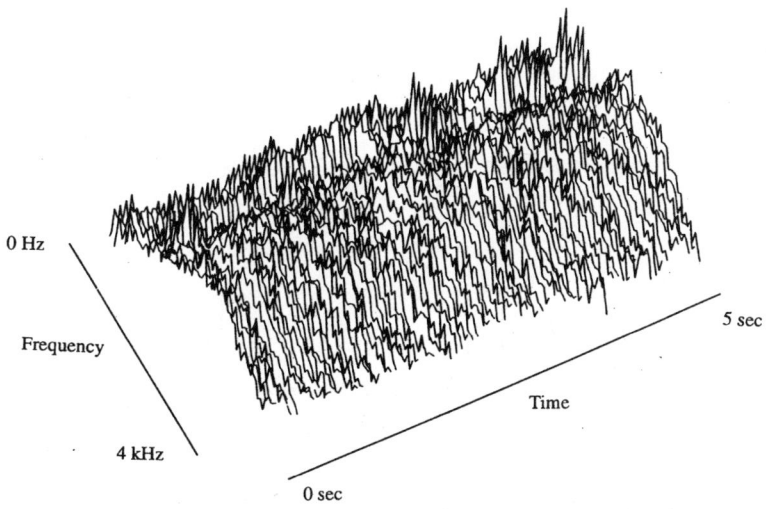

Figure 3.18 A CTFM sonar return that includes a reflection from a metal sphere. Depicted frequency ranges from 0 to 4 kHz over a duration of five seconds.

of these returns relating the maximum peak height in any frequency bin in each time scan to the height of its neighboring bins. These features served as inputs to the neural networks that were evolved to make the proper classification. The total training data set comprised 238 exemplars (101 "mines" and 137 "non-mines"). Figure 3.19 shows a typical learning curve in which the mean squared classification error decreases over successive generations of evolutionary training on a population of 50 neural networks. In the validation of 237 new patterns, the best-evolved network correctly classified 97.03 percent of the "mines" and 96.32 percent of the "non-mines." Porto et al. (1995) provided a follow-up comparison of the evolutionary programming technique with that of simulated annealing and backpropagation on these same data. The results indicated that both evolutionary programming and simulated annealing could outperform backpropagation because it had a tendency to stall in local optima that did not generate networks that generalized well on novel data.

Sebald and Fogel (1990) offered the use of evolutionary programming to train weights in a single-layer neural controller for maintaining the blood pressure of a simulated patient undergoing surgery. This problem had been studied using more conventional neural control methods in Sebald et al. (1989), Karimi and Sebald (1988), and others. The patient's response to a blood pressure control drug (sodium nitroprusside) was modeled as an autoregressive function with an additional polynomial term to account for disturbances in the model:

$$x[t + 1] = \alpha x[t] + \beta u[t - d] + c_0 + tc_1 + t^2c_2 + t^3c_3, \tag{3.24}$$

where $x[t]$ is the blood pressure at time t, $u[t]$ is the input control drug at time t, d is the transport delay for the infusion, and α, β, c_0, c_1, c_2, and c_3 are constants. Reasonable constraints were placed on these values to accommodate knowledge of actual patient response during surgery (e.g., the value of β was constrained between $[-0.1, -1]$). The optimization problem was to evolve the weights of a neural controller that could determine the proper input sequence $\{u\}$ given prior sequence of inputs and patient response $\{x\}$ to maintain the patient's blood pressure at a target

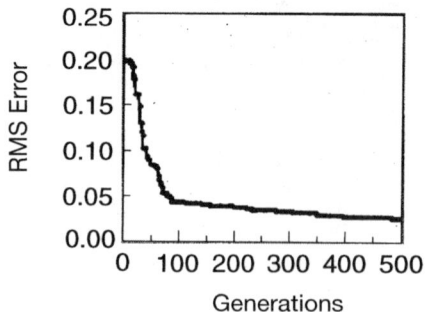

Figure 3.19 A typical rate of optimization when evolving the weights in a neural network to classify the sonar patterns in Porto (1990).

level (typically 70 mm Mg). This was followed up in several subsequent studies (Sebald et al., 1991; Sebald and Schlenzig, 1994; Fogel and Sebald, 1995; and others). The latter efforts used coevolution to produce the models of both the patients and the controllers; that is, a concurrent evolutionary search for the most difficult patients (in one population) was made simultaneously with an evolutionary search for the most suitable controllers (in the other population).

One of the common claims in support of neural networks as pattern classifiers that was heralded in the late 1980s was their assumed fault tolerance. Because the "knowledge" of a network was in some sense distributed across the weights, rather than held in a single place, there was a notion that if some connections in a network were degraded or removed, the overall performance of the network would be generally unaffected (Rumelhart and McClelland, 1986). Unfortunately, experimentation showed that this was not always correct; however, networks could be evolved to be fault tolerant if this were built into the evaluation function. Sebald and Fogel (1992) used a procedure where random weights in a network would be set equal to zero before evaluation in an evolutionary programming optimization of the networks for classifying artifact-free or artifact-laden patterns of heart waves. Over time, networks evolved that could tolerate the random imposition of degraded connections, and they were more robust than other networks trained without having to face such impositions.

Another concern in the design of neural networks is determining the appropriate connectivity between nodes. A completely connected feed forward design may be limited in not allowing recurrent (feedback) connections and may also be overdetermined in that some of the connections may not be necessary. McDonnell and Waagen (1993) used a variety of methods to allow an evolutionary program to devise suitable connectivity for various neural pattern classification problems. One method was based on imposing a penalty function in proportion to the number of connections employed. A bit string was used to indicate which of the possible connections between nodes would be "active." Both the weights and the activity of connections were subject to random variation. Figure 3.20 shows the final evolved single hidden layer network for classifying the T-C problem from Rumelhart et al. (1986), along with the graph indicating the rate of error reduction over successive generations and the number of nodes in the best network at each step. McDonnell and Waagen (1993) also extended the procedure to evolve networks with recurrent connections for time series prediction based on Akaike's information criteria (Akaike, 1974) for penalizing overly complex networks[2] (also see McDonnell and Waagen, 1994).

Angeline et al. (1994) employed an evolutionary program to evolve unconstrained recurrent networks (i.e., any node in the network could be connected to any other with the exception that there could be no links to an input node, no links from an output node, and at most one link between any two specified nodes). Random variation was imposed by altering the weights of networks as well as their structure. When new links were evolved, they were implemented with zero weight so as to attempt to leave the behavior of the network unchanged. Similarly nodes were added

[2]Fogel (1991c) devised a measure related to the AIC for neural network selection.

116 SPECIALIZATIONS

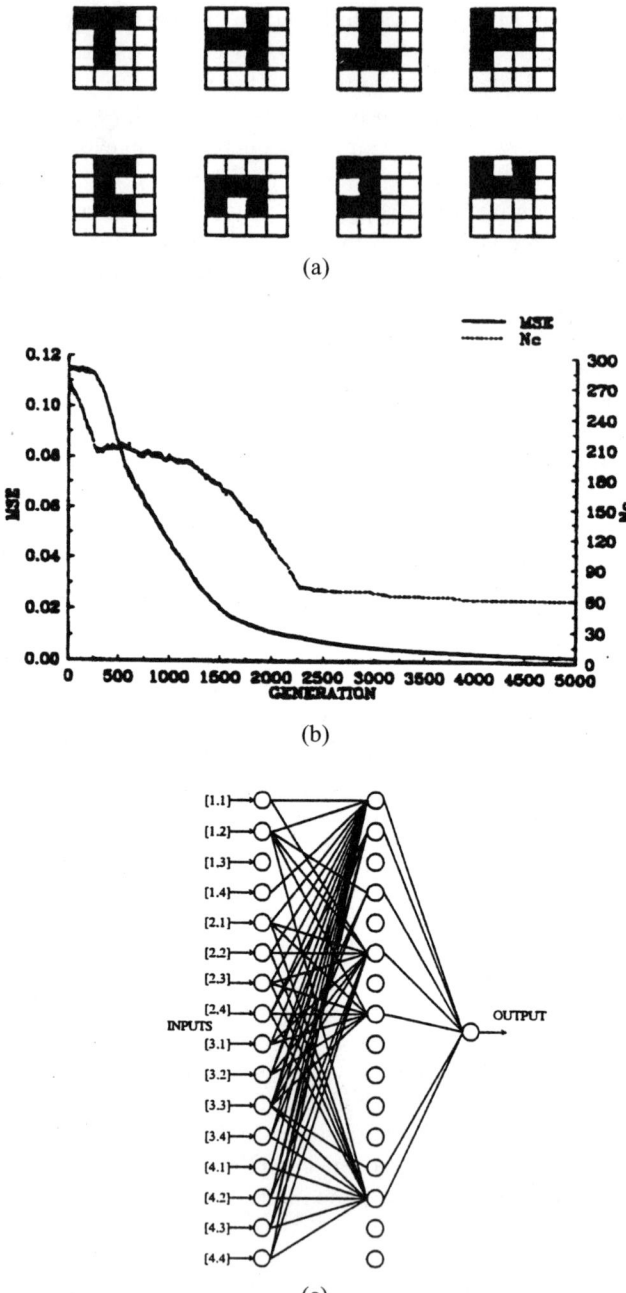

Figure 3.20 (*a*) The T-C problem consists of classifying rotated symbols T or C from a 4 × 4 grid. (*b*) The best-evolved T-C classifier network MSE and number of connections (Nc) at each generation from McDonnell and Waagen (1993). (*c*) The resulting T-C classifier network after 5000 generations. The bias connections are not shown.

unlinked at first. The method was demonstrated on several language induction problems (Tomita, 1982) and the problem of following a trail in a grid ("the ant problem," Jefferson et al., 1992).

Angeline et al. (1994) also argued that recombination was not particularly suitable for optimizing neural architectures. Consider the two networks in Figure 3.21. Each performs the same input-output transfer function (i.e., their behaviors are identical). But if crossover were applied to these networks as parents, their resulting offspring may have very few or none of the behavioral characteristics of either parent. Essentially, the effect of recombination here can be as extreme as a completely

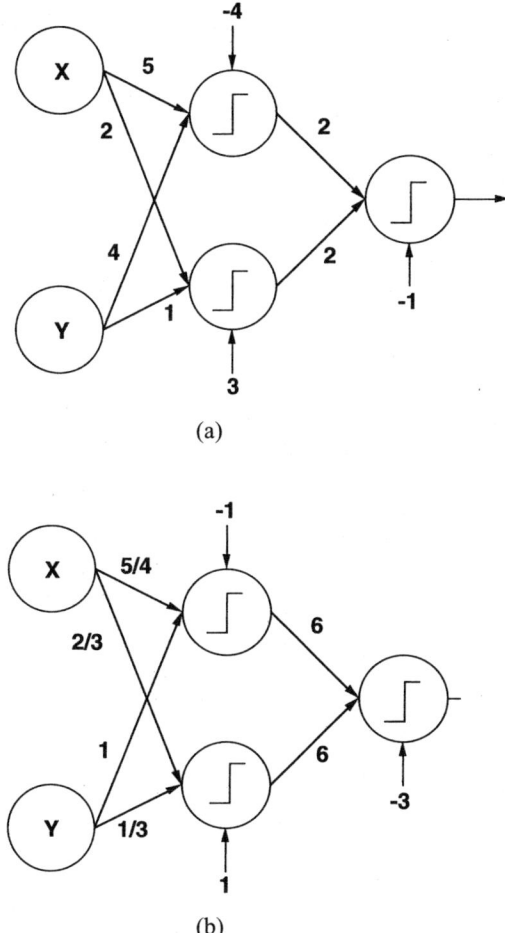

Figure 3.21 Two threshold networks (*a–b*) that provide the same behavior for any input X and Y but have different weights. The threshold functions perform a dot product of the incoming weights and inputs, offset by a bias term, and compare the total to zero. If the result is greater than zero, the threshold is met and the node "fires."

blind search. Constraining crossover to act only at boundaries of nodes in the networks or other higher-order groups of nodes may not solve this problem either because identical networks can be written that are mirror images (see Fig. 3.22), and again, the straightforward application of crossover to such networks does not treat the functionality of each node. Angeline et al. (1994) argued that the use of random variation to all weights in a neural architecture was a generally more appropriate method for retaining the behavioral link between parents and their offspring.

The use of evolutionary programming in optimizing neural architectures has become fairly common with applications in time series prediction (Saravanan, 1993; English, 1994; Yip and Pao, 1994; English and Gotesman, 1995; Yao and Liu, 1997), control of unstable systems (Saravanan and Fogel, 1995), choosing feature sets for training (Brotherton and Simpson, 1995), designing associative memory in Hopfield nets (Imada and Araki, 1997), breast cancer classification (Fogel et al., 1995, 1997, 1998), and the parallel design of networks (Riessen et al., 1997). McDonnell (1992) addressed the issue of optimizing networks in light of constrained weights, but this remains a very open area for research.

Perhaps the greatest effort in this area has been shown by Xin Yao, a professor at the Australian Defence Force Academy in Canberra, Australia. His work has includ-

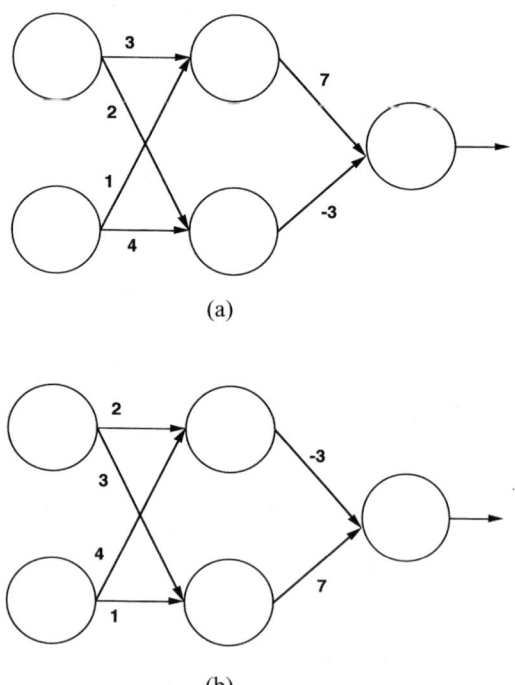

Figure 3.22 Two mirror image neural networks (*a*–*b*). The networks perform the same function because the functionalities of the hidden nodes have simply been swapped.

ed making use of the information in a population of evolving neural classifiers (Yao and Liu, 1998b), evolving both architecture and weights (Yao and Liu, 1998c), evolving modular neural networks (Liu and Yao, 1997), stressing the importance of maintaining the behavioral link between parent networks and their offspring (Yao, 1997), and the design of networks comprising different nodes (Liu and Yao, 1996; this has also been examined more recently in Sebald and Chellapilla, 1998).

3.4 EVOLVING S-EXPRESSIONS AND MULTIPLE INTERACTING PROGRAMS

One area of evolutionary computation that has received recent attention is the design of symbolic expressions that can be interpreted as computer programs. The impetus for this interest came from Koza (1992, 1994) who demonstrated that the essential evolutionary process of population-based random variation and selection could be applied to these structures in order to design programs that solve or approximately solve a variety of engineering problems. In general, the procedure relies on a parse tree representation incorporating functions and terminals (see Fig. 3.23). The prevailing notion of evolving such structures has been to rely on recombination of subtrees as modules or subroutines (i.e., essentially apply a genetic algorithm to this data structure), with the expectation that such crossovers can bring together useful sections of code.

This general method has been demonstrated in a variety of contexts (see Koza et al., 1996, 1997, 1998), but it remains unclear that the crossover of subroutines in such structures is as essential as is often claimed. To investigate this hypothesis, Chellapilla (1997a, 1997b) used an evolutionary programming approach to evolve parse trees to perform many of the same problems offered in Koza (1992). Rather than rely on subtree crossover, a variety of tree-based mutation operations were devised that could be applied to alter a parent structure when creating an offspring (see Fig. 3.24). These operators provide the possibility for a range of change in be-

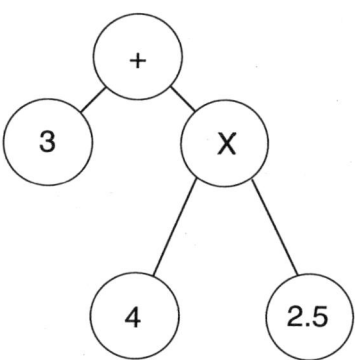

Figure 3.23 Parse tree representation of an expression 3 + (4 ∗ 2.5).

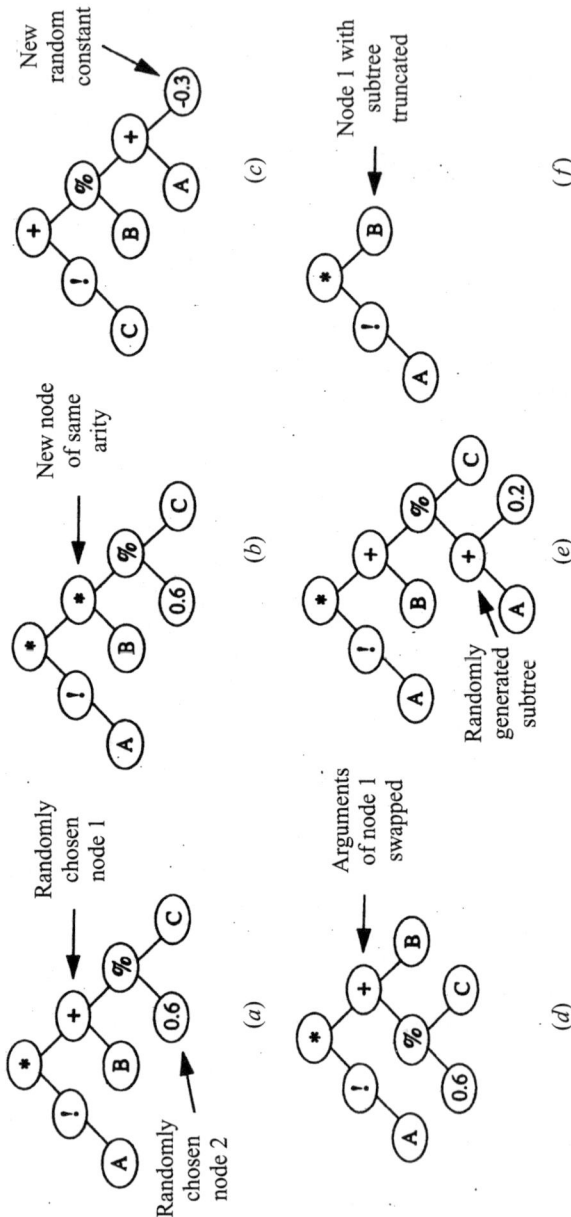

Figure 3.24 Different possible mutations that can be made to a parse tree (from Chellapilla, 1997). Starting with the tree in the upper left-hand corner (*a*), various mutations are shown: (*b*) changing function, (*c*) changing constant value, (*d*) swapping arguments in a subtree, (*e*) generate new subtree, (*f*) truncate subtree.

havior from rather minor to quite significant. In a series of comparisons, Chellapilla (1997a, 1997b) indicated that evolutionary programming could generate appropriate parse trees for these same problems (e.g., Boolean parity, cart centering), often with less expected computational effort and smaller population sizes than appeared necessary when relying on subtree crossover. These results were further pursued with other examples in Chellapilla (1998a, 1998b), involving the control of dynamic systems (e.g., broom balancing and backing up a truck-and-trailer to a designated docking bay) with similar results: The use of designed single-parent variation operators consistently outperformed the results offered in support of subtree recombination. This has now been found to hold even in the case where modules for potentially useful subroutines are designated a priori by the human designer (Chellapilla, 1998c). It is clear that subtree recombination is not required for the successful evolution of programs based on parse trees.

What is not as clear is if the process of subtree crossover actually does what it is believed to do, namely serve as an engine for bringing together building blocks of useful code. Instead of recombining subtrees of existing programs in a population, consider the prospects of recombining an existing parental program with another generated completely at random. Certainly, even though this second method still incorporates the mechanism of crossover, it cannot be considered as usefully recombining building blocks since all that it is actually doing is swapping large sections of random code.[3] This version of recombination is nothing more than a macromutation, and yet results in Angeline (1997a, 1997b) suggest that this macromutation can statistically significantly outperform standard subtree crossover on several problems (e.g., even parity, spiral classification, sunspot prediction). These results are supported by other related experiments offered in Luke and Spector (1997) and O'Reilly (1996). Although there may be cases where recombining subtrees is particularly appropriate, it is seemingly difficult to conjure such cases without having them be contrived.[4]

As is always the case with evolutionary optimization, the success or failure of the procedure depends strongly on the probabilistic behavioral relationship between each parent and its offspring and the degree to which that relationship "fits" the adaptive landscape (response surface) for the problem at hand. No single variation operator should be expected to be any more useful in general than random search (Wolpert and Macready, 1997). Accordingly, the uncritical advocacy of any one particular approach to problem solving that relies on any one specific data structure and variation operator should be avoided.

While a single parse tree can be a useful structure for representing a program, the capabilities of this single tree can be extended by having multiple interacting programs (MIPS) where each parse tree can direct output as input to another program. The end result is very similar to any arbitrarily connected neural network where each node in the net has a potentially arbitrary functionality (see Fig. 3.25). Ange-

---

[3]This was introduced by Jones (1995) with the term "headless chicken crossover" because it employs the mechanism of crossover but is akin to a chicken running around with its head cut off.
[4]The "royal road" problem (Mitchell et al., 1992) is one such problem that was contrived to be easy for a genetic algorithm (but was actually outperformed by a hill climbing strategy).

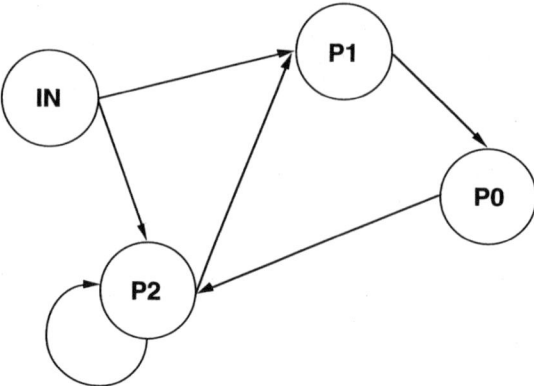

Figure 3.25 A network of multiple interacting programs (MIPS). One node is designated as external input. The nodes P0-P2 contain parse trees that act on external input and the output of other parse trees (and possible previous values of their own output).

line (1997c, 1997d, 1998) has offered preliminary experiments in chaotic time series prediction and the evolution of symbolic equations with this MIPS approach using evolutionary programming to optimize the construction of the interacting programs. Although the ultimate program complexity appears to increase over the use of a single parse tree (possibly greatly), the flexibility of the approach is quite attractive. It remains to be determined if such flexibility is required for current real-world problems.

3.5 GAMES

3.5.1 Iterated Prisoner's Dilemma

A "game" is a mathematical construct involving one or more players (if there is only one player, typically this player is pitted against an environment which takes the place of a second player) in which each can commit allocable resources described as "moves" and each receives a payoff based on the set of moves conducted by all players. The question of which moves to make under what conditions is essentially the question of which *strategy* to employ. It is natural to consider using evolutionary algorithms to optimize such strategies in the face of other players who may be competitive, cooperative, or even neutral. Evolutionary programming has been applied to this end in several regards.

One application involves the well-studied iterated prisoner's dilemma. This is a non-zero-sum, noncooperative game, meaning that the benefits accruing to one player do not necessarily imply similar penalties imposed on the player and that no preplay communication between players is allowed. Collusion is forbidden. The typical prisoner's dilemma provides each player with two alternative moves: cooperate

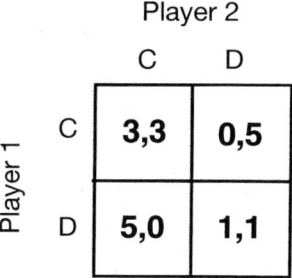

Figure 3.26 The payoff function adopted in the Axelrod prisoner's dilemma tournaments.

(C) or defect (D). Cooperation implies increasing the total gain of both players, while defecting implies increasing one's own reward at the expense of the other player. A typical payoff for each of the four possible outcomes that was used in seminal experiments by Axelrod (1980) is shown in Figure 3.26 (see Axelrod, 1984, for a popular review). As shown, the largest payoff is awarded when one player defects and the other cooperates. The player who defects receives the "temptation" payoff of five units, while the other player receives the "sucker's" payoff of nothing. If both players cooperate, they each receive three units, and if they both defect, then they receive only one unit.

Examining the payoffs indicates an obvious motivation toward mutual cooperation over mutual defection. On the other hand, from the point of view of any single player on any single move, the best play is to defect. If the other player cooperates, defection earns five points instead of just three for cooperating. If the other player instead defects, then defection earns one point instead of nothing for cooperating. And thus we have the "dilemma." For any single play of this game, the best thing to do in a game-theoretic sense is defect. But if the game can be iterated for an unknown number of plays, then there can be evolutionary pressure to cooperate.

These dynamics were first studied in Axelrod (1987) using genetic algorithms to evolve strategies encoded in binary strings.[5] These strategies were limited to a memory of the previous three moves by each player. To alleviate this limitation, Fogel (1991b, 1993b, 1995a) used finite state machines of variable size to represent the strategies adopted by individuals competing for survival in an iterated prisoner's dilemma. To illustrate, one evolved strategy is shown in Figure 3.27. The machine starts in the state indicated by the "start" arrow and first plays the symbol associated with that arrow. After that, its behavior is generated by examining the pair of last moves by both players and the current state. As each new move is elicited, the machine traverses a series of states, and the potential stimulus-response behavior

---

[5]Recall that Fogel and Burgin (1969) used evolutionary programming to explore strategies in the one-shot prisoner's dilemma (see Chapter 2).

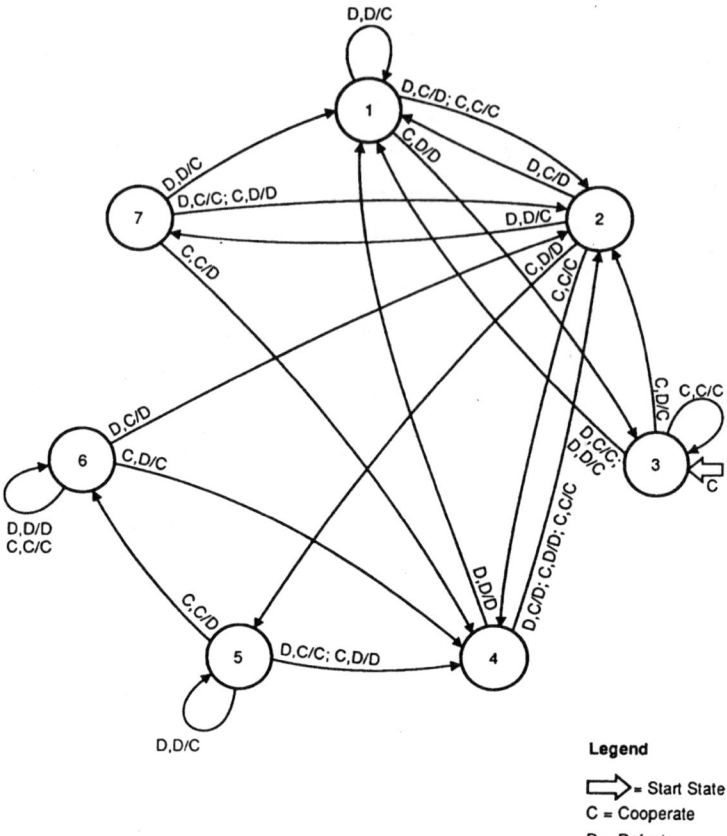

Figure 3.27 A finite state machine representation for a strategy in the iterated prisoner's dilemma (from Fogel, 1993). The machine begins with an intial move and transitions to the start state. The resulting play depends on the pair of previous moves between both players.

changes. Machines of up to eight states were examined, and these can provide a longer memory than was offered in Axelrod (1987).

Fogel (1993c) examined a variety of population sizes and found that when the starting population was initialized completely at random, the resulting emergent behavior was very similar to what was found in Axelrod (1987): The population initially converged toward mutual defection but then quickly rebounded into a long period of mutual cooperation (see Fig. 3.28). That is, cooperation was the reliable longer-term result of the game for populations varying in size from 50 to 1000 parents. Interestingly, however, when surviving machines from different population sizes were played against each other, they often fell into boughts of alternating cooperation and defection, even though they had been cooperative machines when evolved in their original population. Fogel (1993c) speculated that in each evolutionary trial, machines evolved a sequence of initial symbols that is particu-

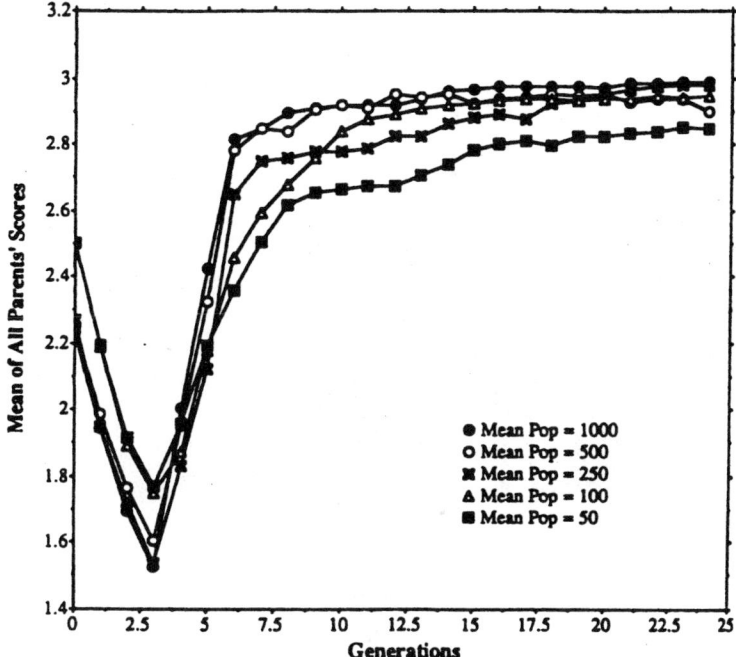

Figure 3.28 The mean fitness for all surviving parents as a function of the number of generations for various population sizes when using evolutionary programming to play the iterated prisoner's dilemma (from Fogel, 1993). There is an initial decline in mean fitness as machines defect early, followed by a subsequent rise toward mutual cooperation.

lar to that trial and represents a "message" that is identified as a signal to cooperate. When machines are removed from separately evolved populations, the intelligibility of this message is lost, and the result can be essentially random. Thus the resulting evolved machines cannot be viewed simply in terms of whether or not they are cooperative but rather under what conditions are they or are they not cooperative.

In each encounter between players in Fogel (1991a, 1993c), the duration of the game was fixed at 151 plays (equivalent to the average of Axelrod's tournaments from the late 1970s). But in the real world, the duration of the encounter can often be limited by one of the players deciding not to play further. Fogel (1995a) examined this possibility by incorporating a "duration length" parameter into each evolving finite state machine. This parameter was subjected to random variation and selection, along with the typical variation and selection applied to the machines (add/delete states, change outputs and state transitions, change the start state or symbol). Interestingly, there was a great variation in the resulting duration of play (Fig. 3.29) with some evidence of a correlation between longer durations and increased mutual cooperation.

Another assumption of the prisoner's dilemma that does not hold in the real

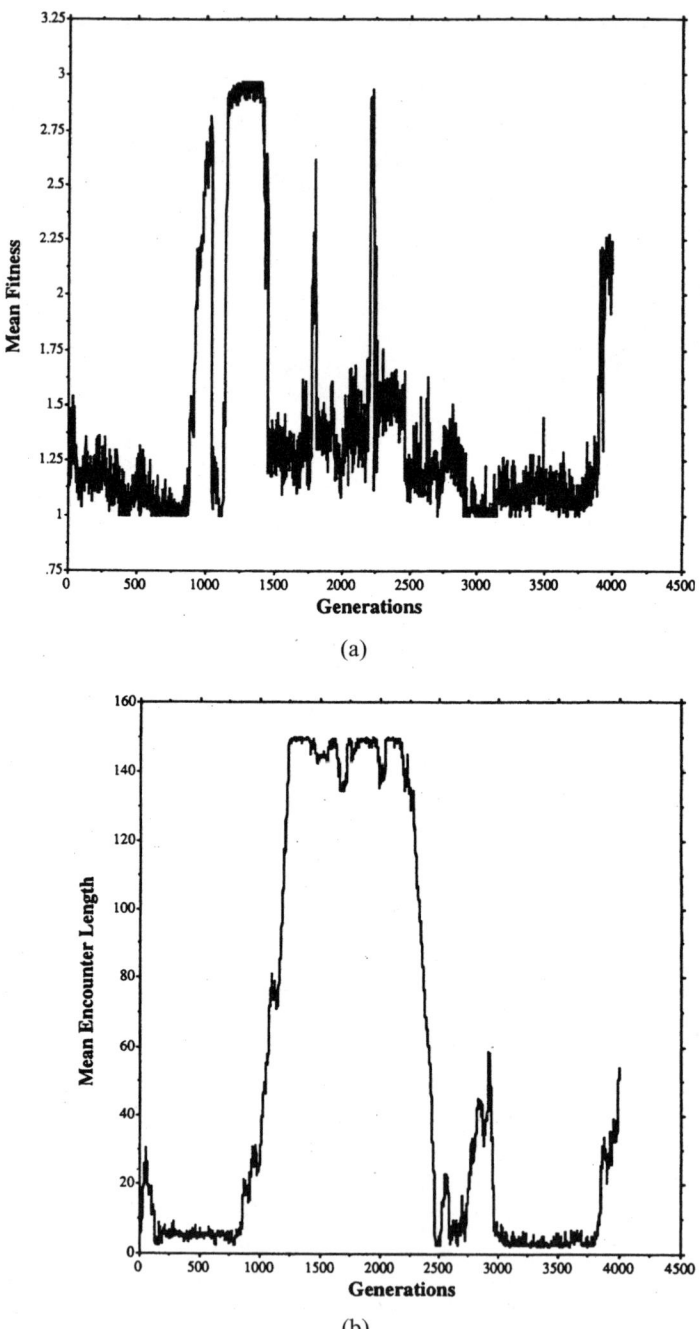

Figure 3.29 For experiments with an evolvable encounter length, (*a*) the mean fitness of all surviving parents and (*b*) the mean encounter length as a function of the number of generations. As encounter lengths grew longer, there was more tendency for mutual cooperation, but this state was not stable (from Fogel, 1995).

world is the dichotomy of cooperate or defect. In natural settings there are degrees of cooperation and defection. Fogel and Harrald (1994) and Harrald and Fogel (1995) explored the possibility for modeling continuous ranges of behavior in the iterated prisoner's dilemma by replacing the finite state machines in Fogel (1993c) with single hidden layer neural networks (Fig. 3.30). The output of the network was constrained to the range [−1, 1], where −1 corresponded to complete defection and +1 corresponded to complete cooperation. A planar approximation to the payoff matrix from Axelrod (1980) was used (Fig. 3.31), and the results were quite different from those observed using just cooperate or defect. For small neural networks (2 hidden nodes) there was a strong tendency to converge on mutual defection, whereas more complex nets (20 hidden nodes) were more likely to generate cooperative behavior. Even then, such cooperative behavior was not stable and tended to degrade over time. In the case of a continuous range of behaviors, cooperation appeared much less tenable.

3.5.2 Learning Tic-Tac-Toe without Knowing the Object of the Game

Imagine playing a game against an expert where you are not told the object of the game and you aren't given any hints about what might be a good or bad move. What's more, consider the case where you don't even get any feedback after a game

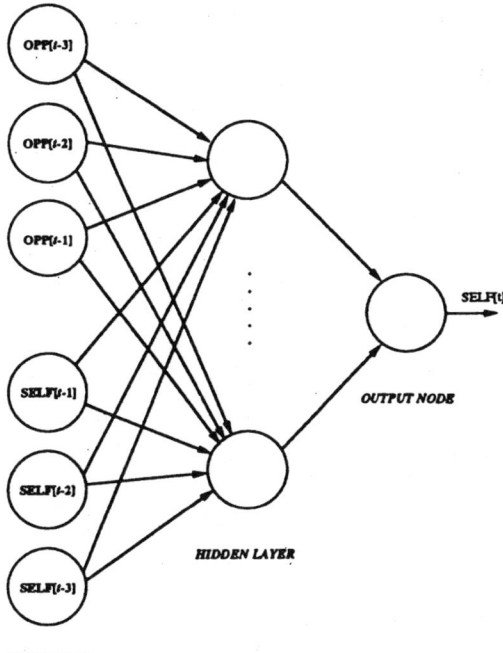

Figure 3.30 Evolving continuous behaviors in the iterated prisoner's dilemma using neural networks (from Harrald and Fogel, 1995). The networks operate on the last three moves and generate output from [−1, 1] ranging from complete defection to complete cooperation.

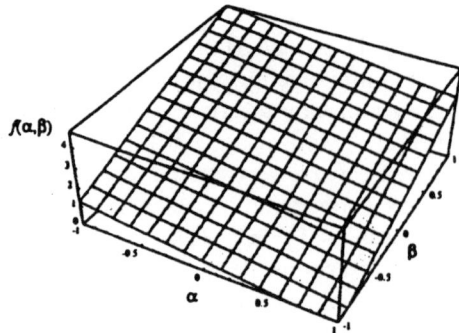

Figure 3.31 The payoff function used in Harrald and Fogel (1995) was a planar approximation to that used in Axelrod (1987).

is completed as to whether or not you won or lost, but instead you are given a point score after playing a series of games so that you don't even know which games were winners and which were losers. This sounds like quite a challenge, and this protocol was taken up by Fogel (1993d, 1995b, pp. 228–243) using evolutionary programming to optimize neural networks to play tic-tac-toe (also known as naughts and crosses, and by several other names).

The game is played in a three-by-three grid and players place markers into the grid in the attempt to get three markers in a row vertically, horizontally, or diagonally, resulting in a "win." Player 1 usually has the marker "X" and player 2 has the marker "O." If a player cannot achieve a win, his secondary objective is to prevent his opponent from winning, that is, playing to a "draw." It is well known that if the second player plays appropriately, he can force at least a draw.

Attention was given to evolving a strategy for the first player when facing an "expert" opponent represented by a rule-based procedure. The player's strategy was represented in a neural network (multilayer perceptron) having nine inputs and outputs, with a variable number of nodes in a single hidden layer. Each hidden node performed the typical dot product of the weighted inputs and passed the result through a sigmoid filter, $(1 + e^{-x})^{-1}$. Each input and output unit corresponded to a square in the grid. The symbols "X" and "O" were mapped to +1 and –1, respectively, with blanks being mapped to 0. A move was determined by presenting the current board pattern (scanned left to right from top to bottom) to the network and examining the relative strengths of the nine output nodes. A marker was placed in the empty square with the maximum output strength (this guaranteed legal moves). The output from nodes associated with squares in which a marker had already been placed was ignored.

The initial population consisted of 50 parent networks. The number of nodes in the hidden layer was chosen at random uniformly over the integers $\{1, \ldots, 10\}$. The initial weighted connection strengths and bias terms were randomly distributed uniformly over the real values [–0.5, 0.5]. A single offspring was copied from each parent and modified by two forms of mutation:

1. All weight and bias terms were perturbed by adding a Gaussian random variable with zero mean and a standard deviation of 0.05.
2. With a probability of 0.5, the number of nodes in the hidden layer was allowed to vary. If a change was indicated, there was an equal likelihood that a node would be added or deleted, subject to the constraints on the maximum and minimum number of nodes (10 and 1, respectively). Nodes to be added were initialized with all weights and the bias term set equal to 0.

A rule-based algorithm that played nearly perfect tic-tac-toe was created to play against the evolving networks. In any particular game, after the neural network played first, the eight possible second moves were stored in an array. The rule base then proceeded as follows:

1. From the array of all possible moves, select a move that has not yet been played.
2. For subsequent moves:
   a. with a 10 percent chance, move randomly, else
   b. if a win is available, place a marker in the winning square, else
   c. if a block is available, place a marker in the blocking square, else
   d. if two open squares are in line with an "O," randomly place a marker in either of the two squares, else
   e. randomly move in any open square.
3. Continue with step 2 until the game is completed.
4. Continue with step 1 until games with all eight possible second moves have been played.

The 10 percent chance of moving randomly generated the possibility for any sequence of moves to be offered by the rule base. Thus it was not enough for the evolving networks learn how to defeat a fixed adversary.

Each network was evaluated over four sets of these eight games. The payoff function varied in three sets of experiments over $\{+1, -1, 0\}$, $\{+1, -10, 0\}$, and $\{+10, -1, 0\}$, where the entries are the payoffs for winning, losing, and playing to a draw, respectively. The maximum possible score over any four sets of games was 32 under the first two payoff functions and 320 under the latter function. But a perfect score in any generation did not necessarily indicate a perfect algorithm because of the random variation that might occur in the rule base's play. After competition against the rule was completed for all networks in the population, the typical probabilistic selection based on a round-robin tournament of size 10 was conducted to select parents for the next generation. Thirty trials were conducted with each payoff function, with each trial being iterated for 800 generations.

Figure 3.32 shows the mean learning rate of the networks when using the $\{+1, -1, 0\}$ payoff function. The 95 percent confidence limits indicate a consistent improvement over increasing generations across all trials. Figure 3.33 shows the tree

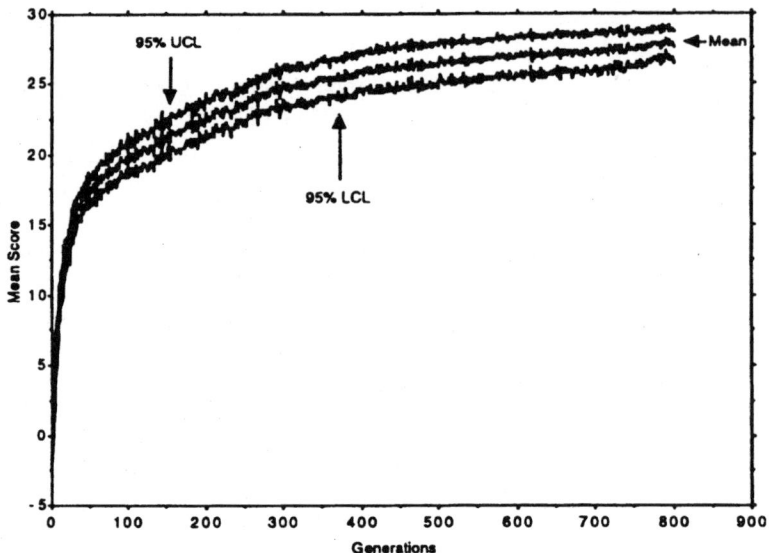

Figure 3.32 The mean and 95% confidence limits on the fitness of the evolving tic-tac-toe player when using a payoff matrix of (+1, −1, 0) for winning, losing, and playing to a draw, respectively (from Fogel, 1995, p. 235). The results are derived from 30 trials.

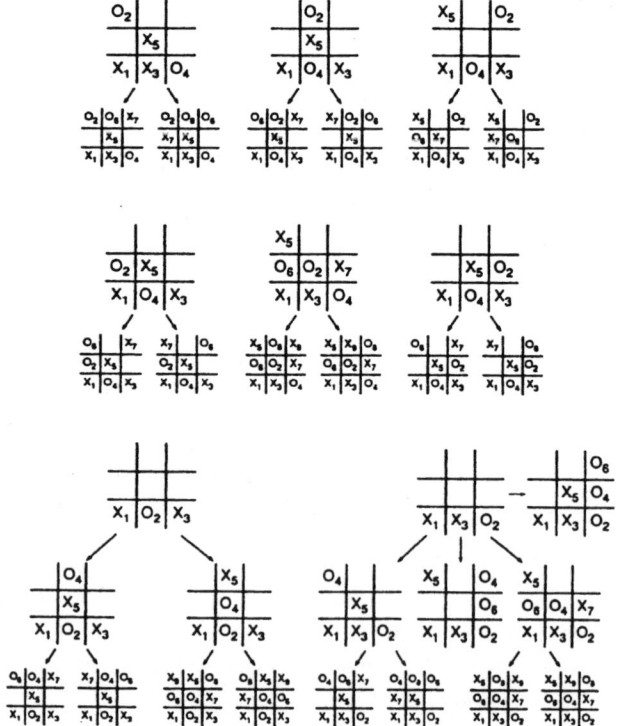

Figure 3.33 The game tree of possible outcomes using the best-evolved network to play against the rule base when using a payoff matrix of (+1, −1, 0). Each of the eight main boards correspond to the eight possible next moves after the neural network moves first. Subscripts indicate the order of play (from Fogel, 1995, p. 237).

of possible games when playing the best-evolved network from the second trial with
{+1, −1, 0} payoffs. The best solution would force a win in four of the eight possible
branches of the tree but might lose in two of the branches. In contrast, Figure 3.34
shows the mean learning rate when using the {+1, −10, 0} payoff function. The initial rate of learning is dramatically faster than before, although the final performance at 800 generations is not as great. Figure 3.35 shows the tree of possible
games with the best network from the first trial: The network never forces a win, but
it also never loses. This reflects the objective given to the networks which penalized
losing at 10 times the reward of winning. Finally Figure 3.36 shows the mean learning rate when using {+10, −1, 0}, which is quite similar to that obtained using {+1,
−1, 0}. Placing greater emphasis on winning had less effect than placing similar emphasis on losing.

Note that the evolved behaviors were adapted to meet the desired goals. The
same evolutionary program learned to play tic-tac-toe in each case without being
told the object of the game, or whether or not it had won or lost in any particular
game. Only a single value was returned after 32 games. No hints were provided, nor
were any mechanisms for evaluating board positions used. Essentially the heuristics
were limited to the facts that there were nine inputs and outputs, and that markers
could only be placed in empty squares. Despite this handicap the evolutionary program was able to evolve networks that were appropriate for the task at hand, although they weren't perfect players. It is interesting to note that Newell (in Minsky,
1961) offered when speaking of checkers or chess: "It is extremely doubtful
whether there is enough information in 'win, lose, or draw' when referred to the
whole play of the game to permit any learning at all over available time scales." The

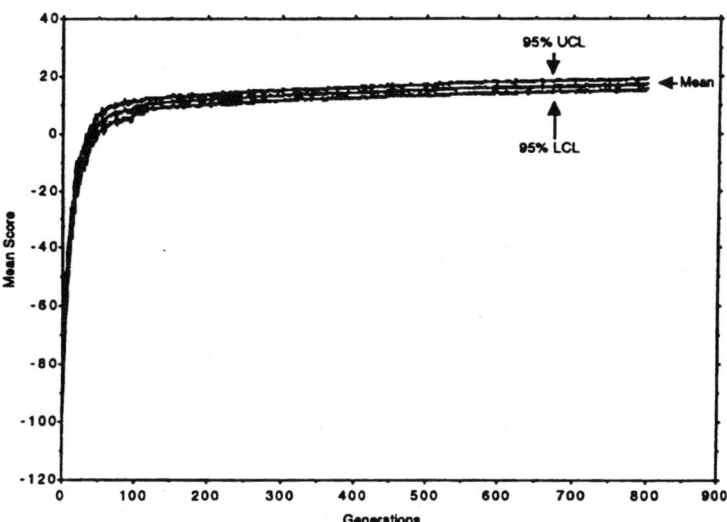

Figure 3.34 The mean and 95% confidence limits on the fitness of the evolving tic-tac-toe
player when using a payoff matrix of (+1, −10, 0) for winning, losing, and playing to a draw,
respectively (from Fogel, 1995, p. 238). The results are derived from 30 trials.

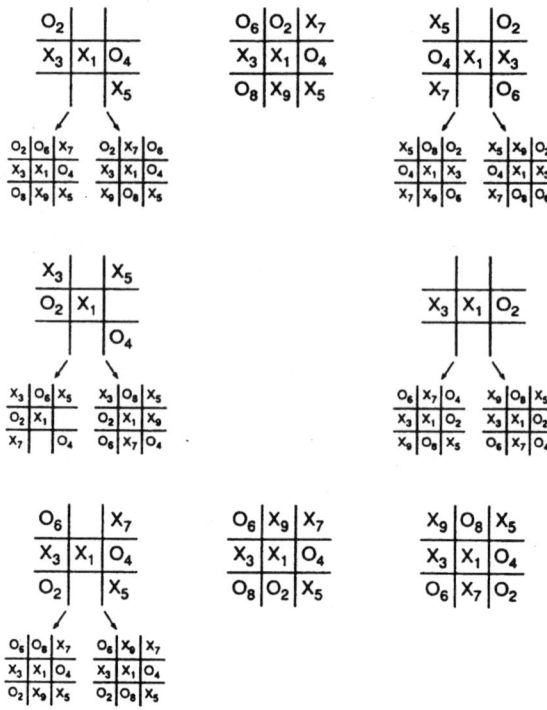

Figure 3.35 The game tree of possible outcomes using the best-evolved network to play again the rule base when using a payoff matrix of (+1, –10, 0) (from Fogel, 1995, p. 240). Note that the network never loses, reflecting the increased penalty for losing.

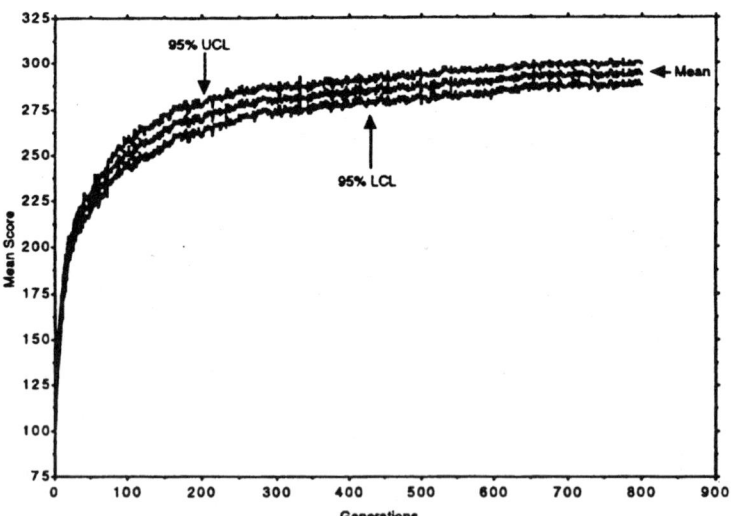

Figure 3.36 The mean and 95% confidence limits on the fitness of the evolving tic-tac-toe player when using a payoff matrix of (+10, –1, 0) for winning, losing, and playing to a draw, respectively (from Fogel, 1995, p. 241). The results are derived from 30 trials and are qualitatively similar to those in Figure 3.32.

veracity of this speculation remains to be determined, but the evidence demonstrates that it does not apply to tic-tac-toe.

3.5.3 Evolving Strategies in Simulated Combat

One of the common examples in gaming involves two or more players in armed conflict. Consider the case of using a simulation that governs the physics of different combat units to train personnel for combat. The simulation can display icons for different vehicles (e.g., tanks) with speed of movement based on terrain, and alternative actions can be carried out in faster than real time. Such "computer-generated forces" (CGFs) make training exercises economically feasible by creating a synthetic, yet realistic battlefield environment. Creating intelligent behaviors for multiple autonomous entities is key to achieving the required realism. Prior research has produced a wide variety of heuristics for controlling forces that operate in a virtual world. Unfortunately, it's often relatively easy to distinguish between human-controlled forces and those controlled by heuristics because the latter are often more predictable. When used for training, subjects learn to defeat the game rather than an intelligently interactive adversary. Training against a force that follows any set of *fixed* rules is inappropriate, for the real enemy learns, demonstrates initiative, and may behave in a truly unpredictable manner.

Realistic simulation of the combat environment must include the opposing force, its decision-making ability, and its mission. Some important steps in this direction have now been offered in Porto and Fogel (1997), where a simulation called ModSAF (modular semi-automated forces) has been adapted to include evolutionary programming. The simulation affords great flexibility and the ability to control a wide variety and multiple number of types of forces. ModSAF emulates low-level behaviors and can be used to manipulate forces at various levels (unit, platoon, etc.), while maintaining physical constraints. By linking evolutionary programming with ModSAF, high-level behaviors can be evolved while relying on the detailed modeling engine to ensure appropriate lower-level behaviors. This realistically simulates a normal chain of command wherein higher-level instructions leave smaller details to be handled in a routine manner.

ModSAF utilizes a taskframe construct wherein a set of sequentially linked taskplans describes the present and future behavior for each simulated entity. These act as a time-ordered set of behavioral plans that specify how, where, when, and what each entity will do in the predicted future. This provides a mechanism for specifying desired (i.e., optimal) behaviors through a given time frame.

ModSAF uses finite state machines to control the behavior of entities in the simulation, such as moving, targeting, and shooting. Some of these are probabilistic in nature; thus there is a chance that a tank firing at a target might not actually hit the target even though it is the predicted and targeted aim point. Multiple simulations of the same scenario are required to acquire sufficient statistical information for reliable estimation of the success or failure rates associated with different behaviors. Essentially the evolutionary program is coupled to ModSAF to evolve alternative finite state behaviors in light of specified goals such as the importance of surviving,

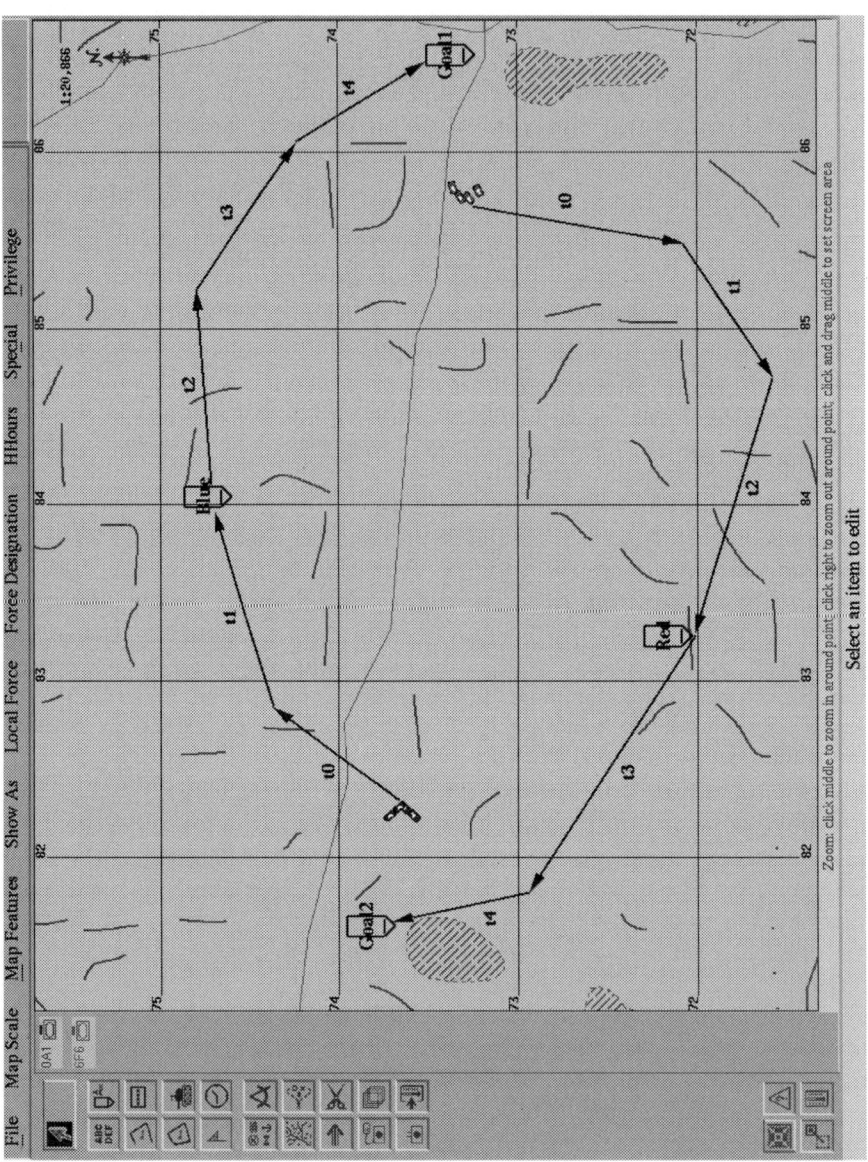

Figure 3.37 Two platoons of tanks, Red and Blue, are assigned similar missions: Reach the goal point while not accepting any risk of being killed. The evolutionary program optimizes the plans for each platoon such that they avoid their mutual firing range. This result was not preprogrammed or rule-based but rather was invented by the evolutionary program on-the-fly.

killing the enemy, arriving at a goal point on time, or minimizing distance traveled. The relative importance of these parameters can be changed on the fly as the simulation is conducted, as desired.

The simulation is conducted in increments of 20 seconds of real time. At each increment an entirely new evolution is conducted to find the next best set of actions based on what can be extrapolated to occur over the next 20 minutes. Opposing forces are projected ahead in time, as are friendly forces, and the resulting situation is evaluated in light of specified criteria. Thus there is a continual adaptation as conditions change and as the mission to be achieved may be altered. Figure 3.37 shows the evolved paths of two platoons of tanks assigned to reach opposing goals, while having a high importance on the probability of survival and a low weight on killing the enemy. Without being instructed using any set of rule-based behaviors, the tanks for both sides evolve paths that keep them out of the firing range of the enemy while progressing to the desired goal. Essentially they have discovered "evasion." Other behaviors which can be described in common terms such as "pursuit," "blockade," and "hunting," have been observed, devised de novo by the evolutionary program.

3.6 OTHER APPLICATIONS

The efforts mentioned so far cover only a small subset of the range of applications of evolutionary programming. A complete description of the totality of these is beyond the scope of this book. But to provide some examples of the variety of these efforts, consider the following list:

- Optimizing fuel distribution to gas stations (McDonnell et al., 1997)
- Medical image analysis (Rizki et al., 1995)
- High-precision control (Kim and Jeon, 1996; Jeon et al., 1997)
- Query translations for multilingual information retrieval (Davis and Dunning, 1995, 1996)
- Predicting protein–protein interactions (Duncan and Olson, 1996)
- Ligand docking (Gehlhaar et al.,1995)
- Determining the hypocenter of earthquakes (Minster et al., 1995)
- Localizing ships using nonacoustic sensors (Porto, 1995)
- Electronic part placement and VLSI design (Nelson, 1995; Prahlada Rao et al., 1995)
- Control of autonomous robots (Kim and Shim, 1995; Yang and Kim, 1996)
- Factory scheduling (Porto, 1997)
- Cutting stock optimization (Liang et al., 1998)
- Fuzzy control (Jeon et al., 1995; Kim and Myung, 1995; Kim and Kim, 1997)
- Optimal trajectory for manipulator (Kim and Kim, 1996a, 1996b)
- Identification and control of systems with friction (Kim et al., 1996)

- Image restoration (Wong and Guan, 1998)
- Optimizing a voltage reference circuit (Nam et al., 1998)
- Various applications in power systems (Lai and Ma, 1996, 1997a, 1997b; Lee and Yan, 1998; Wong and Yuryevich, 1998; Lai, 1998)

In addition several contributions have been made to problems in general engineering such as constrained optimization (Kim and Myung, 1996, 1997; Myung and Kim, 1998), solving finite element problems (Nelson, 1997), ordinary differential equations (Nelson, 1996), generalized mixture densities (Waagen and McDonnell, 1995), scaling evolutionary programming on problems of increasing complexity (Yao and Liu, 1998c), among many others.

3.7 SUMMARY

The above specializations only touch on the different applications and investigations of evolutionary programming since 1990. There are many more that are known but not listed, and undoubtedly others that have been put in practice that have not been published, particularly in the financial and other commercial arenas.

The careful reader will note that in contrast to the many applications and empirical investigations that have been described, there has been relatively little focus on theoretical results in evolutionary programming. This is by design. Although there have been advances in understanding the essential convergence properties of evolutionary programming (e.g., Fogel, 1992a, 1994, showed asymptotic convergence to global optima), these and other related mathematical efforts in other areas of evolutionary computation (e.g., evolution strategies or genetic algorithms) are often not of much practical value.

For example, it is well known that when optimizing on a quadratic bowl (a strongly convex function), a logarithmic speed up in convergence can be obtained by increasing the number of offspring per parent (Bäck et al., 1993, showed this for evolution strategies, and it carries directly over to evolutionary programming). The rate of convergence can be further increased by using an intermediate recombination (averaging) on all parents in the population (Beyer, 1995). But strongly convex surfaces are not characteristic of the problems where evolutionary algorithms can be best put to use—there are other classic algorithms that are superior for these cases. Moreover these types of results can easily be misinterpreted: "Recombination gives faster convergence" sounds like a good thing, but convergence to where . . . ? The answer may be convergence to a local minimum!

While it is important to continue to gain such basic and fundamental understandings of evolutionary programming and other evolutionary algorithms, we must admit that we are a long way from being able to design better practical algorithms as a result of these and other related analyses. Part of the problem is that analyses are too often limited to functions with simplified properties such as convexity or linear separability. A canonical example of the latter is the so-called onemax function in which the object is to find the string of bits such that the sum is maximized with each string being scored based on the number of 1s it contains. The problem is so simple that any attempt to transfer the results of analysis on this function to real-world problems should be viewed with immediate skepticism. Salomon (1996) has shown that the performance of some evolutionary algorithms (e.g., genetic algorithms) often breaks down poorly as components of solutions become less and less linearly separable. But that's the real world!

We must also admit that several previous theoretical speculations in evolutionary computation have proven to be incorrect. For example, Goldberg (1989, p. 80) following Holland (1975) speculated that we should use representations based on minimum cardinality. But the empirical evidence accumulated over the last decade does not support this suggestion. Davis (1991), Michalewicz (1992), Bäck and Schwefel (1993), Fogel and Stayton (1994), and many others, have reported better and faster results in continuous parameter optimization problems when using real-valued representations instead of bit strings. More recently Fogel and Ghozeil (1997) proved that completely equivalent evolutionary algorithms could be constructed regardless of the cardinality of the representation, so the proper conclusion is to use the representation that follows naturally from the problem and gives you the best insight into discovering improved solutions.

And we must recognize that some analysis have simply been incorrect. For example, the formulation in Holland (1975, pp. 75–88) concerning the optimal allocation of trials in the k-armed bandit problem has been shown both by counterexample (Rudolph, 1997) and direct proof (Macready and Wolpert, 1998) to be ill-posed. This seminal analysis lies at the heart of the notion of schema processing and proportional selection in genetic algorithms, and the corrections leave these concepts without substantive theoretical basis. It will take some time for these corrections to become well known, but we cannot shy away from them. It doesn't matter how convincing or well accepted an idea is if it's provably untrue.

While the theoretical results on evolutionary programming (and other evolutionary algorithms) have been hard fought and mostly of little practical value, this has not discouraged their development and its application to real-world problems that have resisted solution by conventional means. The flexibility of the evolutionary approach to problem solving is one of its greatest assets. It allows you to address the real problem at hand, instead of making assumptions that simplify the problem merely for the sake of being able to apply some analytic method—which is a mechanism for generating exactly the right answer to the wrong problem. And the evolutionary approach has the advantage of being adaptable. Rather than having to start from scratch when situations change, you have a population of solutions that have already proven their worth in solving similar problems in the past. The wide range of applications of evolutionary programming offer testament not only to its success but also to its ability to evolve over time. No doubt future generations will continue to evolve the algorithms to solve ever more challenging problems.

OUTLOOK

At the beginning of the 1990s, there were three distinctly different approaches to simulating evolution on a computer. *Genetic algorithms* emphasized binary encodings (bit strings) for representing solutions to problems, which were then modified with a heavy reliance on crossover and minimal reliance on point mutation. Proportional selection was used because of a belief that this would generate an optimal search in a particular mathematical sense. Evolution strategies, in contrast, used continuous parameters for solving engineering problems, with a heavier reliance on mutation and the incorporation of self-adaptation to modify the form of mutation. Selection was stringent, always keeping the best solutions either just from the offspring or from the collection of offspring and their parents. *Evolutionary programming* had explored a variety of variable-length representations and was just starting to incorporate self-adaptation. Selection had become probabilistic, offering the chance for lesser-valued solutions to be maintained as parents, thus broadening the search on the fitness landscape and providing increased chance to escape from local optima.

By the end of the 1990s, the increased communication between these communities resulted in a coalescence of approaches. To consider a few examples, genetic algorithms now use any representation, evolution strategies include greater emphasis on recombination (even as applied to more than two parents at a time), and evolutionary programming has adopted the self-adaptation mechanism offered two decades earlier within evolution strategies. When someone presents a new evolutionary algorithm, it now makes little sense to ask if it's a genetic algorithm, an evolution strategy, or an evolutionary program, for there is so little that separates these approaches. Moreover there is much to be gained by completely eschewing attempts to continue to separate them.

So many of the theoretical justifications for choosing particular representations or selection procedures have been proven to be without mathematical basis (e.g. binary strings, schema processing with proportional selection; see Fogel and Ghozeil, 1997; Rudolph, 1997; Macready and Wolpert, 1998) that we are better directed to thinking about how to design specific evolutionary procedures for specific problems without concern to whether the resulting technique is "genetic" or "evolutionary." And we now know that there cannot be one single best variation operator for all problems. Each method of searching for new solutions can be effective in some instances, but this effectiveness comes at the expense of being ineffective in other instances (Wolpert and Macready, 1997). To overemphasize the importance or specific implementation of any single facet of evolutionary computation is to, by consequence, limit the opportunities for applying the technique.

The essential ingredients of all evolutionary algorithms remain the same: population-based random variation and selection. Further, the critical aspects of these ingredients lie in their interplay. It is infeasible to analyze each aspect in isolation and make "best" choices for population size, or variation operator(s), or selection mechanism because this ignores higher-level interactions. There is no single hallmark to evolutionary algorithms, not crossover, not binary strings, not proportional selection, not self-adaptation. None of these things alone is the keystone to making evolutionary computation successful. The evolutionary approach to problem solving can be best advanced by removing the blinders that keep us focused on specific nuances of the algorithms and instead looking more broadly at how to integrate evolution as value added to routine well-accepted procedures.

Mathematics has long been the primary tool of engineering. Linear approximations yield first-cut solutions. Far more complex representations are required to treat real-world problems. This formalism requires significant computation. We allocate extensive CPU time to solve differential equations that portray time-dependent interactive dynamic systems. But computers can be used more directly. There is no need for the intervening mathematics. The machine itself becomes a fast-time model of natural evolution. Rather than yield a single best design, it provides an evolved population of solutions, each of measured worth. The engineer can review these, taking into account concerns that were not directly stated in the payoff function: A particular design may score well but have undesirable side effects or unintentional consequences. Conventional or current designs and plans can be introduced into the population as suggestions. If they are of true worth, they will be so recognized by having survived the selection process. Evolutionary computation is not in competition with other methods of optimization; it simply provides a way of improving on their results. In essence it is a simulation of the scientific method and therefore an embodiment of creativity.

But there is more to do. We need to determine the fundamental capabilities and limitations of evolutionary computation. What is the simplest way to generate useful heredity through reproduction in a finite noisy environment? What form of evolutionary computation should be used to treat which class of problems? What are the trade-offs between population size, number of generations, type of variation, and so forth? Are there ways to estimate the convergence rate under different conditions, particularly when facing multimodal problems?

It might also prove worthwhile to examine the manner in which evolutionary computation relates to natural evolution. Is it worthwhile to replicate a specific variation and selection process that takes place in nature? How should genetic variation be combined with phenotypic selection? Is it worthwhile to replicate natural evolution across the hierarchy from genes through organisms (their organization and species)? What theory of natural evolution (Lamarckian, Darwinian, and so forth) is most appropriate for addressing certain kinds of problems? Would it be advantageous to combine aspects of these different theories?

How can evolutionary computation be made still more efficient and effective in particular cases? Is it worthwhile to use asynchronous evolution to allow some of the better parents to reproduce across generations and/or to artificially alter the response surface (the adaptive landscape) to facilitate finding the global optimum? How can evolutionary computation benefit from alternative forms of self-adaptation (internally and/or through the use of meta-levels)? What are the limits of evolving self-referential, self-modeling, self-aware, conscious automata that, in an artificial social setting, may even exhibit conscience, the prerequisite for accepting responsibility?

How should evolutionary computation be realized under different conditions/constraints? When and how should distributed/parallel processing be used for evolutionary computation? When is it advantageous to hardwire evolutionary computation into a chip? These and other questions are worthy of serious consideration.

Evolutionary computation is a broadly useful tool for finding solutions to difficult problems. But that presupposes that you know what problem you face. Unfortunately, all too often in the course of assisting others to solve their problems, I have found that problems are often left undefined, the purpose to be achieved is only poorly stated.

The following scenario occurs commonly in widely different settings. The person in charge calls together his advisors, indicates the problem, then asks for advice—for plans, policies, tactics, strategies, courses of action—they go by different names depending on the context, but they are all alternative ways to solve the problem, to meet the need. Each advisor offers one or more plans that reflect their particular understanding of the purpose to be achieved. One advisor thinks the person in charge wants a high-risk solution, another believes a low-risk solution would be preferable. But the person in charge never mentioned the acceptable level of risk, or risk of what! Another advisor thinks he only wants a near-term solution, while yet another believes he is looking for a long-term plan. Yet he never indicated the time frame of concern. In point of fact, each plan was conceived only in terms of the individual advisor's own understanding of the purpose to be achieved. And differences in viewpoint concerning the purpose to be achieved may be significant at all levels within an organization.

Even specific engineering problems often remain ill-defined. The requirements are stated. But are these hard or soft constraints? A constraint is hard when it must be met or else the measurable worth of the proposed solution is zero. If the constraints are soft, how soft? What else should be measured? Are these measures equally important? Are some, or all of them, critical (that means failure in this regard nullifies the value of any other achievement)? Are there degrees of criticality, and if so, how should these be taken into account? We often simply "minimize mean squared error" rather than refer to the real payoff matrix wherein equally correct outcomes are not of equal worth and equal but opposite errors have different cost. We often refer to measures of effectiveness (MOEs) and measures of performance (MOPs), but clearly these alone do not tell the whole story.

Effective management begins with a clear understanding of what must be achieved, by when. But more often than not, organizations publish "vision statements" that indicate only their most desired future. What if that outcome cannot be fully realized? Surely some value is found in lesser degrees of achievement. There are even times when our primary concern is to avoid a particularly undesirable situation. In other words, to be meaningful, a statement of purpose must be a statement of the relative worth of each of the significantly different futures, ranging all the way from utopia to catastrophe. Only then can we measure the overall worth of *any* given situation.

Defining these significantly different futures can be a considerable challenge. Rather than directly face this difficulty, I have devised an alternative approach where the individual aspects of concern across the various futures are listed. These form the dimensions of a hyperspace. Each parameter is attributed some relative importance and made measurable by designating mutually exclusive class intervals that indicate those differences that make a difference in degree of achievement on each of these parameters. The class intervals thus defined must exhaustively span the range from the most to the least desirable degree of achievement, each being attributed some value for that degree of achievement.

Parameters are often posed that cannot be measured directly. It is then useful to express these parameters in terms of subsidiary measures (subparameters) that can be made measurable. If there is difficulty in this regard, it is useful to use still lower-level subparameters. This procedure is extended until all the dimensions of concern are directly measurable. Thus purpose often takes the form of a hierarchic *Valuated State Space*. A normalizing or combinatory function is then specified so that the contribution on the various parameters and subparameters can be properly aggregated into a single overall worth for any given situation. The Valuated State Space and normalizing function then constitute the required multiattribute utility function.

It is worthwhile to note that there has been considerable academic interest in such multiobjective optimization problems. One typical approach is to attempt to discover the so-called Pareto optimal set that contains all solutions such that any change to a particular parameter results in a worse solution. But this approach is often insufficient because there are intrinsic differences between solutions in the Pareto set. If I present a manager with different options from this set, they will very likely identify some as being better than others; what is most unlikely is that they will agree that these are all of equal value! Searching for Pareto optimal solutions is only a first step. Finding out how to make the essential discrimination between alternative solutions is what's required and what can be accomplished using the Valuated

State Space approach. Properly defining the purpose to be achieved is prerequisite for success in using evolutionary computation.

On May 7, 1998, at the World Congress on Computational Intelligence in Anchorage, Alaska, I was honored with the first IEEE Neural Networks Council Pioneer Award in Evolutionary Computation and was elected a Fellow of the IEEE. The wording on my plaque reads: "For the invention of evolutionary computation." This is of course an overstatement. There were many others who participated in the "invention" of these algorithms. Many came before me, like Hans Bremermann, George Box, Alex Fraser, Richard Friedberg, George Friedman, and Nils Barricelli. Others came later, like Ingo Rechenberg, Hans-Paul Schwefel, John Holland, and Michael Conrad. But those who came later were no less pioneers than those who came earlier. We were all converging toward a central idea of simulating the essential aspects of evolution for optimization, adaptation, or artificial intelligence. We were all working independently, without the benefit of the means for rapid dissemination of information like email and the internet. And we were all doing something a bit different, different enough to be generally dismissed out of hand as trying something that "would never fly."

In a real sense, however, if you are working in evolutionary computation right now, then you are a pioneer as well. It has only been 45 years since Barricelli published his first paper on artificial life based on the work he did in John von Neumann's lab at Princeton. Half a century is all the time that has transpired in evolutionary computation. As compared to other disciplines, like mathematics, physics, chemistry, or biology, or even computer science, evolutionary computation is in its infancy. What is now required is logic and experiment. In the words of Galileo, "In questions of science, the authority of thousands is not worth the humble reasoning of a single individual," and in the words of Cole Porter (1933):

> Experiment, Make it your motto day and night.
> Experiment, And it will lead you to the light.
> The apple on top of the tree is never
> too high to achieve.
> So take an example from Eve,
> EXPERIMENT, Be curious
> Tho' interfering friends may frown,
> Get furious at each attempt to hold you down.
> If this advice you only employ,
> The future can offer you infinite joy and merriment,
> EXPERIMENT, and you'll see.

REFERENCES

H. Akaike (1974), A new look at the statistical model identification, *IEEE Trans. Autom. Control* 19: 716–723.

B. L. Andersen, W. C. Page, and J. R. McDonnell (1991), Multi-output system identification using evolutionary programming, *Proc. 25th Asilomar Conf. on Signals, Systems, and Computers,* R. R. Chen (ed.), Los Alamitos, CA: IEEE Computer Society, pp. 546–550.

B. K. Ambati, J. Ambati, and M. M. Mokhtar (1991), Heuristic combinatorial optimization by simulated Darwinian evolution: A polynomial time algorithm for the traveling salesman problem, *Biolog. Cybern.* 65: 31–35.

P. B. Anderson and D. W. Dearholt (1974), A program for automata evolution, Computer Science Conf., Detroit, MI, February.

P. J. Angeline (1996), The effects of noise on self-adaptive evolutionary optimization, *Evolutionary Programming V: Proc. 5th Ann. Conf. on Evolutionary Programming,* L. J. Fogel, P. J. Angeline, and T. Bäck (eds.), Cambridge: MIT Press, pp. 433–439.

P. J. Angeline (1997a), Subtree crossover: Building block engine or macromutation? *Genetic Programming 1997: Proc. 2nd Ann. Conf.,* J. R. Koza, K. Deb, M. Dorigo, D. B. Fogel, M. Garzon, H. Iba, and R. L. Riolo (eds.), San Francisco: Morgan Kaufmann, pp. 9–17.

P. J. Angeline (1997b), Comparing subtree crossover with macromutation, *Evolutionary Programming VI: Proc. 6th Ann. Conf. on Evolutionary Programming,* P. J. Angeline, R. G. Reynolds, J. R. McDonnell, and R. Eberhart (eds.), Berlin: Springer, pp. 101–112.

P. J. Angeline (1997c), The benefits of distributed solutions when evolving symbolic equations, *Applications of Soft Computing,* B. Bosacchi, J. C. Bezdek, and D. B. Fogel (eds.), SPIE vol. 3165, San Diego, CA, pp. 124–134

P. J. Angeline (1997d), An alternative to indexed memory for evolving programs with explicit state representation, *Genetic Programming 1997: Proc. 2nd Ann. Conf. on Genetic Programming,* J. R. Koza, K. Deb, M. Dorigo, D. B. Fogel, M. Garzon, H. Iba, and R. L. Riolo (eds.), San Francisco: Morgan Kaufmann, pp. 423–430.

P. J. Angeline (1998), Evolving predictors for chaotic time series, *Applications and Science of Computational Intelligence,* S. K. Rogers, D. B. Fogel, J. C. Bezdek, and B. Bosacchi (eds.), SPIE vol. 3390, Orlando, FL, pp. 170–180.

P. J. Angeline, D. B. Fogel, and L. J. Fogel (1996), A comparison of self-adaptation methods for finite state machines in a dynamic environment, *Evolutionary Programming V,* L. J. Fogel, P. J. Angeline, and T. Bäck (eds.), Cambridge: MIT Press, pp. 441–449.

P. J. Angeline, G. M. Saunders, and J. B. Pollack (1994), An evolutionary algorithm that constructs recurrent neural networks, *IEEE Trans. Neural Net.* 5: 54–65.

J. W. Atmar (1976), Speculation on the evolution of intelligence and its possible realization in machine form, Doctoral dissertation, New Mexico State University, Las Cruces.

R. Axelrod (1980), Effective choice in the prisoner's dilemma, *J. Conflict Resol.,* 24: 3–25.

R. Axelrod (1984), *The Evolution of Cooperation,* New York: Basic Books.

R. Axelrod (1987), The evolution of strategies in the iterated prisoner's dilemma, *Genetic Algorithms and Simulated Annealing,* L. Davis (ed.), London: Pitman, pp. 32–41.

T. Bäck (1992), personal communication, Univ. Dortmund, Germany.

T. Bäck, G. Rudolph, and H.-P. Schwefel (1993), Evolutionary programming and evolution strategies: Similarities and differences, *Proc. 2nd Ann. Conf. on Evolutionary Programming,* D. B. Fogel and W. Atmar (eds.), La Jolla, CA: Evolutionary Programming Society, pp. 11–22.

T. Bäck and H.-P. Schwefel (1993), An overview of evolutionary algorithms for parameter optimization, *Evol. Comp.* 1: 1–24.

H.-G. Beyer (1995), Towards a theory of evolution strategies: On the benefit of sex—the $(\mu/\mu, \lambda)$-theory, *Evol. Comp.* 3: 81–111.

I. O. Bohachevsky, M. E. Johnson, and M. L. Stein (1986), Generalized simulated annealing for function optimization, *Technometrics* 28: 209–218.

E. Bonomi and J.-L. Lutton (1984), The *n*-city traveling salesman problem: Statistical mechanics and the metropolis algorithm, *SIAM Rev.* 26: 551–568.

C. L. Bridges and D. E. Goldberg (1987), An analysis of reproduction and crossover in a binary-coded genetic algorithm, *Genetic Algorithms and Their Applications: Proc. 2nd Int. Conf. on Genetic Algorithms,* J. J. Grefenstette (ed.), Hillsdale, NJ: Lawrence Erlbaum, pp. 9–13.

T. W. Brotherton and P. K. Simpson (1995), Dynamic feature set training of neural nets for classification, *Evolutionary Programming IV: Proc. 4th Ann. Conf. on Evolutionary Programming,* J. R. McDonnell, R. G. Reynolds, and D. B. Fogel (eds.), Cambridge: MIT Press, pp. 83–94.

G. H. Burgin (1969), On playing two-person zero-sum games against nonminimax players, *IEEE Trans. Syst. Sci. Cybern.* 5: 369–370.

G. H. Burgin (1974), System identification by quasilinearization and evolutionary programming, *J. Cybern.,* 2: 4–23.

K. Chellapilla (1997a), Evolutionary programming with tree mutations: Evolving computer programs without crossover, *Genetic Programming 1997: Proc. 2nd Ann. Conf. on Genetic Programming,* J. R. Koza, K. Deb, M. Dorigo, D. B. Fogel, M. Garzon, H. Iba, and R. L. Riolo. (eds.), Cambridge: MIT Press, pp. 431–438.

K. Chellapilla (1997b), Evolving computer programs without subtree crossover, *IEEE Trans. Evol. Comp.* 1: 209–216.

K. Chellapilla (1998a), Automatic generation of nonlinear optimal control laws for broom balancing using evolutionary programming, *Proc. 1998 IEEE Int. Conf. on Evolutionary Computation,* Anchorage, AK, pp. 195–200.

K. Chellapilla (1998b), Evolving nonlinear controllers for backing up a truck-and-trailer using evolutionary programming, *Evolutionary Programming VII: Proc. 7th Ann. Conf. on Evolutionary Programming,* V. W. Porto, N. Saravanan, D. E. Waagen, and A. E. Eiben (eds.), Berlin: Springer, pp. 417–426.

K. Chellapilla (1998c), A preliminary investigation into evolving modular programs without subtree crossover, *Genetic Programming 1998: Proc. 3rd Ann. Conf. on Genetic Programming,* J. R. Koza, W. Banzhaf, K. Chellapilla, K. Deb, M. Dorigo, D. B. Fogel, M. H. Garzon, D. E. Goldberg, H. Iba, and R. L. Riolo (eds.), San Francisco: Morgan Kaufmann, pp. 23–31.

K. Chellapilla and D. B. Fogel (1997a), Exploring self-adaptive methods to improve the efficiency of generating approximate solutions to traveling salesman problems using evolutionary programming, *Evolutionary Programming VI: Proc. 6th Ann. Conf. on Evolutionary Programming,* P. J. Angeline, R. G. Reynolds, J. R. McDonnell, and R. Eberhart (eds.), Berlin: Springer, pp. 361–371.

K. Chellapilla and D. B. Fogel (1997b), Two new mutation operators for enhanced search and optimization in evolutionary programming, *Applications of Soft Computing,* B. Bosacchi, J. C. Bezdek, and D. B. Fogel (eds.), SPIE vol. 3165, pp. 260–269.

A. C. Clarke (1962), From Lilliput to Brobdingnag, *Playboy Magazine,* 9: 102,116–119.

F. N. Cornett (1972), An application of evolutionary programming to pattern recognition, MS thesis, New Mexico State University, Las Cruces.

F. N. Cornett, V. P. Holmes, and D. W. Dearholt (1973), Some experiments with evolutionary programs, *Proc. 7th Ann. Conf. on Information Sciences and Systems,* March.

C.-J. Chung and R. G. Reynolds (1997), Function optimization using evolutionary programming with self-adaptive cultural algorithms, *Simulated Evolution and Learning 1996,* X. Yao, J.-H. Kim, and T. Furuhashi (eds.), Berlin: Springer, pp. 17–26,.

L. Davis (ed.) (1987), *Genetic Algorithms and Simulated Annealing,* London: Pitman.

L. Davis (ed.) (1991), *Handbook of Genetic Algorithms,* New York: Van Nostrand Reinhold.

M. W. Davis and T. E. Dunning (1995), Query translation using evolutionary programming for multilingual information retrieval, *Evolutionary Programming IV: Proc. 4th Ann. Conf. on Evolutionary Programming,* J. R. McDonnell, R. G. Reynolds, and D. B. Fogel (eds.), Cambridge: MIT Press, pp. 175–185.

M. W. Davis and T. E. Dunning (1996), Query translation using evolutionary programming for multilingual information retrieval II, *Evolutionary Programming V: Proc. 5th Ann. Conf. on Evolutionary Programming,* L. J. Fogel, P. J. Angeline, and T. Bäck (eds.), Cambridge: MIT Press, pp. 103–112.

M. W. Davis (1994), The natural formation of Gaussian mutation strategies in evolutionary programming, *Proc. 3rd Ann. Conf. on Evolutionary Programming,* A. V. Sebald and L. J. Fogel (eds.), Singapore: World Scientific, pp. 242–252.

L. Davis and M. Steenstrup (1987), Genetic algorithms and simulated annealing: An overview, *Genetic Algorithms and Simulated Annealing,* L. Davis (ed.), London: Pitman, pp. 1–11.

D. W. Dearholt (1976), Some experiments on generalization using evolving automata, *Proc. 9th Hawaii Intern. Conf. on System Sciences 1976,* Honolulu: Western Periodicals, pp. 131–133.

K. A. De Jong, W. M. Spears, and D. F. Gordon (1995), Using Markov chains to analyze GAFOs, *Foundations of Genetic Algorithms 3,* L. D. Whitley and M. D. Vose (eds.), San Mateo, CA: Morgan Kaufmann, pp. 115–137.

B. S. Duncan and A. J. Olson (1996), Applications of evolutionary programming for the prediction of protein-protein interactions, *Evolutionary Programming V: Proc. 5th Ann. Conf. on Evolutionary Programming,* L. J. Fogel, P. J. Angeline, and T. Bäck (eds.), Cambridge: MIT Press, pp. 411–417.

T. M. English (1994), Generalization in populations of recurrent neural networks, *Proc. 3rd Ann. Conf. on Evolutionary Programming,* A. V. Sebald and L. J. Fogel (eds.), Singapore: World Scientific, pp. 26–33.

T. M. English and M. Gotesman (1995), Stacked generalization and fitness ranking in evolutionary algorithms, *Evolutionary Programming IV: Proc. 4th Ann. Conf. on Evolutionary Programming,* J. R. McDonnell, R. G. Reynolds, and D. B. Fogel (eds.), Cambridge: MIT Press, pp. 205–218.

L. J. Eshelman and J. D. Schaffer (1995), Crossover's niche, *Proc. 5th Intern. Conf. on Genetic Algorithms,* S. Forrest (ed.), San Mateo, CA: Morgan Kaufmann, pp. 9–14.

E. A. Feigenbaum (1963), Artificial intelligence research, *IEEE Trans. Inf. Theory,* 9: 248–253.

M. M. Flood (1962), Stochastic learning theory applied to choice experiments with cats, dogs, and men, *Behavioral Sci.* 7: 289–314.

D. B. Fogel (1988), An evolutionary approach to the traveling salesman problem, *Biolog. Cybern.* 60: 139–144.

D. B. Fogel (1990a), A parallel processing approach to a multiple traveling salesman problem using evolutionary programming, *Proc. 4th Ann. Symp. on Parallel Processing,* Fullerton, IEEE Computer Society, pp. 318–326.

D. B. Fogel (1990b), Evolutionary system identification and control, *Proc. IECON'90: 16th Ann. Conf. of IEEE Industrial Electronics Society,* A. C. Weaver (chair), Piscataway, NJ: IEEE Press, pp. 1271–1274.

D. B. Fogel (1991a), The evolution of intelligent decision making in gaming, *Cybern. Syst.* 22: 223–236.

D. B. Fogel (1991b), *System Identification through Simulated Evolution: A Machine Learning Approach to Modeling,* Needham, MA: Ginn.

D. B. Fogel (1991c), An information criterion for optimal neural network selection, *IEEE Trans. Neural Net.* 2: 490–497.

D. B. Fogel (1992a), Evolving artificial intelligence, Doctoral dissertation, University of California at San Diego, La Jolla.

D. B. Fogel (1992b), Using evolutionary programming for modeling: An ocean acoustic example, *IEEE J. Ocean Eng.* 17: 333–340.

D. B. Fogel (1993a), Applying evolutionary programming to selected traveling salesman problems, *Cybern. Syst.* 24: 27–36.

D. B. Fogel (1993b), Empirical estimation of the computation required to discover approximate solutions to the traveling salesman problem using evolutionary programming, *Proc. 2nd Ann. Conf. on Evolutionary Programming,* D. B. Fogel and W. Atmar (eds.), La Jolla, CA: Evolutionary Programming Society, pp. 56–61.

D. B. Fogel (1993c), Evolving behaviors in the iterated prisoner's dilemma, *Evol. Comp.* 1: 77–97.

D. B. Fogel (1993d), Using evolutionary programming to create neural networks that are capable of playing tic-tac-toe, *Proc. 1993 Int. Conf. on Neural Networks,* Piscataway, NJ: IEEE Press, pp. 875–880.

D. B. Fogel (1994), Asymptotic convergence properties of genetic algorithms and evolutionary programming: Analysis and experiments, *Cybern. Syst.* 25: 389–407.

D. B. Fogel (1995a), On the relationship between the duration of an encounter and the evolution of cooperation in the iterated prisoner's dilemma, *Evol. Comp.* 3: 349–363.

D. B. Fogel (1995b), *Evolutionary Computation: Toward a New Philosophy of Machine Intelligence,* Piscataway, NJ: IEEE Press.

D. B. Fogel (ed.), (1998a), *Evolutionary Computation: The Fossil Record,* Piscataway, NJ: IEEE Press.

D. B. Fogel (1998b), personal communication, Natural Selection, Inc., La Jolla, CA.

D. B. Fogel and J. W. Atmar (1990), Comparing genetic operators with Gaussian mutations in simulated evolutionary processes using linear systems, *Biolog. Cybern.* 63: 111–114.

D. B. Fogel and K. Chellapilla (1998), Revisiting evolutionary programming, SPIE Aerosense98, *Applications and Science of Computational Intelligence,* S. K. Rogers, D. B. Fogel, J. C. Bezdek, and B. Bosacchi (chairs), Orlando, FL, pp. 2–11

D. B. Fogel and L. J. Fogel (1990), Optimal routing of multiple autonomous underwater vehicles through evolutionary programming, *Proc. Symp. Autonomous Underwater Vehicle Technology,* C. Stuart (chair), Washington DC, pp. 44–47.

D. B. Fogel and L. J. Fogel (1996), Preliminary experiments on discriminating between chaotic signals and noise using evolutionary programming, *Genetic Programming 1996: Proc. 1st Ann. Conf. on Genetic Programming,* J. R. Koza, D. E. Goldberg, D. B. Fogel, and R. L. Riolo (eds.), Cambridge: MIT Press, pp. 512–520.

D. B. Fogel, L. J. Fogel, and J. W. Atmar (1991), Meta-evolutionary programming, *Proc. 25th Asilomar Conf. on Signals, Systems, and Computers,* R. R. Chen (ed.), Los Alamitos, CA: IEEE Computer Society, pp. 540–545.

D. B. Fogel, L. J. Fogel, W. Atmar, and G. B. Fogel (1992), Hierarchic methods of evolutionary programming, *Proc. 1st Ann. Conf. on Evolutionary Programming,* D. B. Fogel and W. Atmar (eds.), La Jolla, CA: Evolutionary Programming Society, pp. 175–182.

D. B. Fogel, L. J. Fogel, and V. W. Porto (1990a), Evolutionary programming for training neural networks, *Proc. 1990 Int. Joint Conf. on Neural Networks,* vol. 1, Piscataway, NJ: IEEE Press, pp. 601–605.

D. B. Fogel, L. J. Fogel, and V. W. Porto (1990b), Evolving neural networks, *Biolog. Cybern.,* 63: 487–493.

D. B. Fogel and A. Ghozeil (1997), A note on representations and variation operators, *IEEE Trans. Evol. Comp.* 1: 159–161.

D. B. Fogel and A. V. Sebald (1995), Steps toward controlling blood pressure during surgery using evolutionary programming, *Evolutionary Programming IV: Proc. 4th Ann. Conf. on Evolutionary Programming,* J. R. McDonnell, R. G. Reynolds, and D. B. Fogel (eds.), Cambridge: MIT Press, pp. 69–82.

D. B. Fogel and P. G. Harrald (1994), Evolving continuous behaviors in the iterated prisoner's dilemma, *Proc. 3rd Ann. Conf. on Evolutionary Programming,* A. V. Sebald and L. J. Fogel (eds.), River Edge, NJ: World Scientific, pp. 119–130.

D. B. Fogel and P. K. Simpson (1993a), Evolving fuzzy clusters, *Proc. Int. Conf. on Neural Networks 1993,* San Francisco: IEEE Press, pp. 1829–1834.

D. B. Fogel and P. K. Simpson (1993b), Experiments with evolving fuzzy clusters, *Proc. 2nd Annual Conf. on Evolutionary Programming,* D. B. Fogel and W. Atmar (eds.), La Jolla, CA: Evolutionary Programming Society, pp. 90–97.

D. B. Fogel and L. C. Stayton (1994), On the effectiveness of crossover in simulated evolutionary optimization, *BioSyst.* 32: 171–182.

D. B. Fogel, E. C. Wasson, and E. M. Boughton (1995), Evolving neural networks for detecting breast cancer, *Cancer Lett.* 96: 49–53.

D. B. Fogel, E. C. Wasson, E. M. Boughton, and V. W. Porto (1997), A step toward computer-assisted mammography using evolutionary programming and neural networks, *Cancer Lett.* 119: 93–97.

D. B. Fogel, E. C. Wasson, E. M. Boughton, and V. W. Porto (1998), Evolving artificial neural networks for screening features from mammograms, *Artificial Intelligence in Medicine.* 14: 317–326.

L. J. Fogel (1962a), Autonomous automata, *Indus. Res.* 4: 14–19.

L. J. Fogel (1963), *Biotechnology: Concepts and Applications,* Englewood, NJ: Prentice-Hall.

L. J. Fogel (1964), On the organization of intellect, Doctoral dissertation, University of California at Los Angeles.

L. J. Fogel (1990), The future of evolutionary programming, *Proc. 24th Asilomar Conf. on Signals, Systems, and Computers,* R. R. Chen (ed.), Pacific Grove, CA: Maple Press, pp. 1036–1038.

L. J. Fogel, P. J. Angeline, and D. B. Fogel (1994), A preliminary investigation on extending evolutionary programming to include self-adaptation on finite state machines, *Informatica* 18: 387–398.

L. J. Fogel, P. J. Angeline, and D. B. Fogel (1995), An evolutionary programming approach to self-adaptation in finite state machines, *Evolutionary Programming IV: Proc. 4th Ann. Conf. on Evolutionary Programming,* J. R. McDonnell, R. G. Reynolds, and D. B. Fogel (eds.), Cambridge: MIT Press, pp. 355–365.

L. J. Fogel and G. H. Burgin (1969), Competitive goal-seeking through evolutionary programming, Air Force Cambridge Research Labs, Final report, Contract AF 19(628)-5927.

L. J. Fogel and D. B. Fogel (1986), Artificial intelligence through evolutionary programming, Final report, U. S. Army Research Institute PO-9–X56–1102C-1, Titan Systems, Inc., San Diego, CA, October.

L. J. Fogel and R. A. Moore (1968), Modeling the human operator with finite-state machines, Final report, NASA CR-1112, Decision Science, Inc., San Diego, CA, July.

L. J. Fogel, A. J. Owens, and M. J. Walsh (1964), On the evolution of artificial intelligence, *Proc. 5th National Symp. on Human Factors in Engineering,* IEEE, San Diego, CA, pp. 63–76.

L. J. Fogel, A. J. Owens, and M. J. Walsh (1965a), Intelligent decision-making through a simulation of evolution, *IEEE Trans. Hum. Factors Electron.* 6: 13–23.

L. J. Fogel, A. J. Owens, and M. J. Walsh (1965b), Intelligent decision making through a simulation of evolution, *Behavioral Sci.* 11: 253–272.

L. J. Fogel, A. J. Owens, and M. J. Walsh (1965c), Artificial intelligence through a simulation of evolution, *Biophysics and Cybernetic Systems: Proc. 2nd Cybernetic Sciences Symp.,* M. Maxfield, A. Callahan, and L. J. Fogel (eds.), Washington DC: Spartan Books, pp. 131–155.

L. J. Fogel, A. J. Owens, and M. J. Walsh (1966), *Artificial Intelligence through Simulated Evolution,* New York: Wiley.

D. K. Gehlhaar and D. B. Fogel (1996), Tuning evolutionary programming for conformationally flexible molecular docking. *Evolutionary Programming V: Proc. 5th Ann. Conf. on Evolutionary Programming,* L. J. Fogel, P. J. Angeline, and T. Bäck (eds.), Cambridge: MIT Press, pp. 419–429.

D. K. Gehlhaar, G. Verkhivker, P. A. Rejto, D. B. Fogel, L. J. Fogel, and S. T. Freer (1995), Docking conformationally flexible small molecules into a protein binding site through evolutionary programming, *Evolutionary Programming IV: Proc. 4th Ann. Conf. on Evolutionary Programming,* J. R. McDonnell, R. G. Reynolds, and D. B. Fogel (eds.), Cambridge: MIT Press, pp. 615–627.

A. Ghozeil and D. B. Fogel (1996a), Discovering patterns in spatial data using evolutionary programming, *Genetic Programming 1996: Proc. 1st Ann. Conf. on Genetic Programming,* J. R. Koza, D. E. Goldberg, D. B. Fogel, and R. L. Riolo (eds.), Cambridge: MIT Press, pp. 521–527.

A. Ghozeil and D. B. Fogel (1996b), A preliminary investigation into directed mutations in evolutionary algorithms, *Parallel Problem Solving from Nature 4,* H.-M. Voigt, W. Ebeling, I. Rechenberg, and H.-P. Schwefel (eds.), Berlin: Springer, pp. 329–335.

D. E. Goldberg (1985), Genetic algorithms and rule learning in dynamic system control, *Proc. Int. Conf. on Genetic Algorithms and Their Applications,* J. J. Grefenstette (ed.), Hillsdale, NJ: Lawrence Erlbaum, pp. 8–15.

D. E. Goldberg (1989), *Genetic Algorithms in Search, Optimization and Machine Learning,* Reading, MA: Addison-Wesley.

D. E. Goldberg and R. Lingle (1985), Alleles, loci, and the traveling salesman problem, *Proc. Int. Conf. on Genetic Algorithms and Their Applications,* J. J. Grefenstette (ed.), Hillsdale, NJ: Lawrence Erlbaum, pp. 154–159.

J. J. Grefenstette (ed.), (1985), *Proc. Int. Conf. on Genetic Algorithms and Their Applications,* Hillsdale, NJ: Lawrence Erlbaum.

J. J. Grefenstette (ed.), (1987), *Genetic Algorithms and Their Applications: Proc. 2nd Int. Conf. on Genetic Algorithms,* Hillsdale, NJ: Lawrence Erlbaum.

G. Gunther (1962), Cybernetic ontology and transjunctional operations, *Self-organizing Systems,* M. C. Yovits, G. T. Jacobs, and G. D. Goldstein (eds.), Washington DC: Spartan Books, pp. 313–392.

P. G. Harrald and D. B. Fogel (1995), Evolving continuous behaviors in the iterated prisoner's dilemma, *BioSyst.* 37: 135–145.

G. H. Hardy and E. M. Wright (1962), *An Introduction to the Theory of Numbers,* Oxford: Oxford University Press.

J. H. Holland (1975), *Adaptation in Natural and Artificial Systems,* Ann Arbor: University of Michigan Press.

V. P. Holmes (1973), Recognizing prime numbers with an evolutionary program, MS thesis, New Mexico State University, Las Cruces.

A. Imada and K. Araki (1997), Searching real-valued synaptic weights of Hopfield's associative memory using evolutionary programming, *Evolutionary Programming VI,* P. J. Angeline, R. G. Reynolds, J. R. McDonnell, and R. Eberhart (eds.), Berlin: Springer, pp. 13–22.

D. Jefferson, R. Collins, C. Cooper, M. Dyer, M. Flowers, R. Korf, C. Taylor, and A. Wang (1992), The genesys/tracker system, *Artificial Life II,* C. G. Langton, C. Taylor, J. D. Farmer, and S. Rasmussen (eds.), Reading, MA: Addison-Wesley, pp. 549–577.

J.-Y. Jeon, J.-H. Kim, and K. Koh (1997), Experimental evolutionary programming-based high-precision control, *IEEE Control Syst. Technol.* 17: 66–74.

T. Jones (1995), Crossover, macromutation, and population-based search, *Proc. 6th Int. Conf. on Genetic Algorithms,* L. Eshelman (ed.), Palo Alto, CA: Morgan Kaufmann, pp. 73–80.

A. Karimi and A. V. Sebald (1988), Computer-aided design of closed-loop controllers for biomedical applications, *Proc. 10th Ann. Int. Conf. IEEE Eng. Medicine and Biology Society,* pp. 1404–1405.

J.-H. Kim and Shim (1995), Evolutionary programming-based optimal robust locomotion control of autonomous mobile robots, *Evolutionary Programming IV: Proc. 4th Ann. Conf. on Evolutionary Programming,* J. R. McDonnell, R. G. Reynolds, and D. B. Fogel (eds.), Cambridge: MIT Press, pp. 631–644.

J.-H. Kim and H. Myung (1995), Fuzzy logic control using evolutionary programming and principle of maximum entropy, *Proc. 1st Int. Symp. on Fuzzy Logic (ISFL),* Switzerland, pp. C122–C127.

J.-H. Kim and H. Myung (1996), A two-phase evolutionary programming for general constrained optimization problem, *Evolutionary Programming V: Proc. 5th Ann. Conf. on Evolutionary Programming,* L. J. Fogel, P. J. Angeline, and T. Bäck (eds.), Cambridge: MIT Press, pp. 295–304.

J.-H. Kim and H. Myung (1997), Evolutionary programming techniques for constrained optimization problems, *IEEE Trans. Evol. Comp.* 1: 129–140.

J.-H. Kim, H.-K. Chae, J.-Y. Jeon, and S.-W. Lee (1996), Identification and control of systems with friction using accelerated evolutionary programming, *IEEE Control Syst. Technol.* 16: 38–47.

J.-H. Kim and J.-Y. Jeon (1996), Evolutionary programming-based high-precision control design, *Evolutionary Programming V: Proc. 5th Ann. Conf. on Evolutionary Programming,* L. J. Fogel, P. J. Angeline, and T. Bäck (eds.), Cambridge: MIT Press, pp. 73–81.

J.-H. Kim and K.-C. Kim (1997), Multicriteria fuzzy control using evolutionary programming, *Inf. Sci. Appl.,* February.

S. Kim and J.-H. Kim (1996a), Optimal trajectory planning of a redundant manipulator using evolutionary programming, *Proc. 1996 IEEE Int. Conf. on Evolutionary Computation,* Nagoya, Japan, pp. 738–743.

S. Kim and J.-H. Kim (1996b), Near-global optimal trajectory planning of a redundant manipulator using evolutionary computation, *Proc. IEEE Int. Conf. on Industrial Electronics, Control and Instrumenation (IECON),* pp. 1909–1914.

S. Kirkpatrick, C. D. Gelatt, and M. P. Vecchi (1983), Optimization by simulated annealing, *Science* 220: 671–680.

J. R. Koza (1992), *Genetic Programming,* Cambridge: MIT Press.

J. R. Koza (1994), *Genetic Programming II,* Cambridge: MIT Press.

J. R. Koza, D. E. Goldberg, D. B. Fogel, and R. L. Riolo (eds.), (1996), *Genetic Programming 1996: Proc. 1st Ann. Conf. on Genetic Programming,* Cambridge: MIT Press.

J. R. Koza, K. Deb, M. Dorigo, D. B. Fogel, M. Garzon, H. Iba, and R. L. Riolo (eds.), (1997), *Genetic Programming 1997: Proc. 2nd. Ann. Conf. on Genetic Programming,* San Francisco: Morgan Kaufmann.

J. R. Koza, W. Banzhaf, K. Chellapilla, K. Deb, M. Dorigo, D. B. Fogel, M. H. Garzon, D. E. Goldberg, H. Iba, and R. L. Riolo (eds.), (1998), *Genetic Programming 1998: Proc. 3rd Ann. Conf. on Genetic Programming,* San Francisco: Morgan Kaufmann.

L. L. Lai (1998), *Intelligent System Applications in Power Engineering: Evolutionary Programming and Neural Networks,* New York: Wiley.

L. L. Lai and J. T. Ma (1996), Application of evolutionary programming to transient and subtransient parameter estimation, *IEEE Trans. Energy Convers.* 11: 523–529.

L. L. Lai and J. T. Ma (1997a), Application of evolutionary programming to reactive power planning—Comparison with nonlinear programming, *IEEE Trans. Power Syst.* 12: 198–204.

L. L. Lai and J. T. Ma (1997b), New approach of using evolutionary programming to reactive power planning with network contingencies, *European Trans. Electr. Power* 7: 211–216.

K. Y. Lee and F. F. Yang (1998), Optimal reactive power planning using evolutionary algorithms: A comparative study for evolutionary programming, evolutionary strategy, genetic algorithm and linear programming, *IEEE Trans. Power Syst.* 13: 101–108.

D. B. Lenat (1983), The role of heuristics in learning by discovery: Three case studies, *Machine Learning: An Artificial Intelligence Approach,* R. S. Michalski, J. G. Carbonell, and T. M. Mitchell (eds.), Palo Alto, CA: Tioga Publishing, pp. 243–306.

K.-H. Liang, X. Yao, C. Newton, and D. Hoffman (1998), Solving cutting stock problems by evolutionary programming, *Evolutionary Programming VII: Proc. 7th Ann. Conf. on Evolutionary Programming,* V. W. Porto, N. Saravanan, D. E. Waagen, and A. E. Eiben (eds.), Berlin: Springer, pp. 755–764.

R. K. Lindsay (1968), Artificial evolution of intelligence, *Comtemp. Psych.* 13: 113–116.

Y. Liu and X. Yao (1996), Evolutionary design of artificial neural networks with different nodes, *Proc. 1996 IEEE International Conference on Evolutionary Computation,* Nagoya, Japan, Piscataway, NJ: IEEE Press, pp. 670–675.

Y. Liu and X. Yao (1997), Evolving modular neural networks which generalise well, *Proc. 1997 IEEE Int. Conf. on Evolutionary Computation,* Indianapolis, IN: IEEE Press, pp. 605–610.

R. D. Luce and H. Raiffa (1957), *Games and Decisions,* New York: Wiley.

S. Luke and L. Spector (1997), A comparison of crossover and mutation in genetic programming, *Genetic Programming 1997: Proc. 2nd Ann. Conf. on Genetic Programming,* J. R. Koza, K. Deb, M. Dorigo, D. B. Fogel, M. Garzon, H. Iba, and R. L. Riolo, (eds.), San Francisco: Morgan Kaufmann, pp. 240–248.

B. E. Lutter (1968), The application of artificial intelligence through the evolutionary programming technique to the control of chemical engineering processes, MS thesis, South Dakota School of Mines and Technology, Rapid City.

B. E. Lutter and R. C. Huntsinger (1969), Engineering applications of finite automata, *Simulation* 13: 5–11.

B. E. Lutter and J. P. O'Connor (1967), Digital computer application of a graphical solution to cooling tower design, *Proc. South Dakota Academy of Sciences,* vol. 46, p. 260.

M. R. Lyle (1972), An investigation into scoring techniques in evolutionary programming, MS thesis, New Mexico State University, Las Cruces.

W. G. Macready and D. H. Wolpert (1998), Bandit problems and the exploration/exploitation trade off, *IEEE Trans. Evol. Comp.* 2: 2–22.

J. R. McDonnell (1992), Training neural networks with weight constraints, *Proc. 1st Ann. Conf. on Evolutionary Programming,* D. B. Fogel and W. Atmar (eds.), La Jolla, CA: Evolutionary Programming Society, pp. 111–119.

J. R. McDonnell and D. Waagen (1993), Neural network structure design by evolutionary programming, *Proc. 2nd Ann. Conf. on Evolutionary Programming,* D. B. Fogel and W. Atmar (eds.), La Jolla, CA: Evolutionary Programming Society, pp. 79–89.

J. R. McDonnell and D. Waagen (1994), Evolving recurrent perceptrons for time-series modeling, *IEEE Trans. Neural Net.* 5: 24–38.

J. R. McDonnell, W. C. Page, D. B. Fogel, and L. J. Fogel (1997), Optimizing fuel distribution through evolutionary programming, *Evolutionary Programming VI: Proc. 6th Ann. Conf. on Evolutionary Programming,* P. J. Angeline, R. G. Reynolds, J. R. McDonnell, and R. Eberhart (eds.), Berlin: Springer, pp. 373–382.

Z. Michalewicz (1992), *Genetic Algorithms + Data Structures = Evolution Programs,* Berlin: Springer.

D. Michie (1970), Future for integrated cognitive systems, *Nature* 228: 717–722.

M. L. Minksy (1961), Steps toward artificial intelligence, *Proc. IRE* 49: 8–30.

M. L. Minsky and S. Papert (1969), *Perceptrons: An Introduction to Computational Geometry,* Cambridge: MIT Press.

J.-B. H. Minster, N. P. Williams, T. G. Masters, J. F. Gilbert, and J. S. Haase (1995), Application of evolutionary programming to earthquake hypocenter determination, *Evolutionary Programming IV: Proc. 4th Ann. Conf. on Evolutionary Programming,* J. R. McDonnell, R. G. Reynolds, and D. B. Fogel (eds.), Cambridge: MIT Press, pp. 3–17.

M. Mitchell, S. Forrest, and J. H. Holland (1992), The royal road for genetic algorithms: Fitness landscapes and GA performance, *Proc. 2st Europ. Conf. on Art. Life: Towards a Practice of Autonomous Systems,* F. J. Varela and P. Bourgine (eds.), Cambridge: MIT Press, pp. 245–254.

J. Montez (1974), Evolving automata for classifying electrocardiograms, MS thesis, New Mexico State University, Las Cruces.

O. Morgenstern (1966), On some criticisms of game theory, in *Theory of Games, Techniques and Applications,* A. Mensh (ed.), New York: Elsevier, pp. 446–447.

H. Myung and J.-H. Kim (1998), Hybrid interior-Lagrangian penalty based evolutionary optimization, *Evolutionary Programming VII: Proc. 7th Ann. Conf. on Evolutionary Programming,* V. W. Porto, N. Saravanan, D. E. Waagen, and A. E. Eiben (eds.), Berlin: Springer.

D. Nam, Y. D. Seo, L.-J. Park, C. H. Park, and B. Kim (1998), Parameter optimization of a voltage reference circuit using EP, *Proc. 1998 IEEE Int. Conf. on Evolutionary Computation,* Piscataway, NJ: IEEE Press, pp. 301–305.

J. F. Nash (1954), *Non-cooperative Games: Annals of Mathematics Studies,* no. 54, Princeton: Princeton University Press.

K. M. Nelson (1995), A comparison of evolutionary programming and genetic algorithms for electronic part placement, *Evolutionary Programming IV: Proc. of 4th Ann. Conf. on Evolutionary Programming,* J. R. McDonnell, R. G. Reynolds, and D. B. Fogel (eds.), Cambridge: MIT Press, pp. 503–519.

K. M. Nelson (1996), Solution methods for ordinary differential equations using evolutionary algorithms, *Evolutionary Programming V: Proc. 5th Ann. Conf. on Evolutionary Programming,* L. J. Fogel, P. J. Angeline, and T. Bäck (eds.), Cambridge: MIT Press, pp. 131–138.

K. M. Nelson (1997), Using evolutionary programming for finite element problems, *Evolutionary Programming VI: Proc. 6th Ann. Conf. on Evolutionary Programming,* P. J. Angeline, R. G. Reynolds, J. R. McDonnell, and R. Eberhart (eds.), Berlin: Springer, pp. 397–406

D. J. Nettleton and R. Garigliano (1994), Evolutionary algorithms and a fractal inverse problem, *BioSyst.*, 33: 221–232.

U.-M. O'Reilly (1996), Investigating the generality of automatically defined functions, *Genetic Programming 1996: Proc. 1st Ann. Conf. on Genetic Programming*, J. R. Koza, D. E. Goldberg, D. B. Fogel, and R. L. Riolo (eds.), Cambridge: MIT Press, pp. 351–356.

V. W. Porto (1989), Detection of undersea objects using neural networks, *Proc. 23rd Asilomar Conf. on Signals, Systems, and Computers*, vol. 1, R. R. Chen (ed.), San Jose, CA: Maple Press, pp. 376–380.

V. W. Porto (1990), Evolutionary methods for training neural networks for underwater pattern classification, *Proc. 24th Asilomar Conf. on Signals, Systems, and Computers*, R. R. Chen (ed.), San Jose, CA: Maple Press, pp. 1015–1019.

V. W. Porto (1995), Non-acoustic sensor array localization using evolutionary programming, *Evolutionary Programming V: Proc. 5th Ann. Conf. on Evolutionary Programming*, L. J. Fogel, P. J. Angeline, and T. Bäck (eds.), Cambridge: MIT Press, pp. 19–32.

V. W. Porto (1997), On the practical application of evolutionary programming to optimize job shop scheduling, *IPMM'97: Australiasia-Pacific Forum on Intelligence Processing and Manufacturing of Materials*, vol. 1, T. Chandra, S. R. Leclair, J. A. Meech, B. Verma, M. Smith, and B. Balachandran (eds.), Gold Coast, Australia, pp. 593–599.

V. W. Porto, D. B. Fogel, and L. J. Fogel (1995), Alternative neural network training methods, *IEEE Expert*, June, pp. 16–22.

V. W. Porto and L. J. Fogel (1997), Evolution of intelligently interactive behaviors for simulated forces, *Evolutionary Programming VI*, P. J. Angeline, R. G. Reynolds, J. R. McDonnell, and R. Eberhart (eds.), Berlin: Springer, pp. 419–430.

V. W. Porto, N. Saravanan, D. E. Waagen, and A. E. Eiben (eds.), (1998), *Evolutionary Programming VII: Proc. 7th Ann. Conf. on Evolutionary Programming*, Berlin: Springer.

B. B. Prahlada Rao, L. M. Patnaik, and R. C. Hansdah (1995), An extended evolutionary programming algorithm for VLSI channel routing, *Evolutionary Programming IV: Proc. of 4th Ann. Conf. on Evolutionary Programming*, J. R. McDonnell, R. G. Reynolds, and D. B. Fogel (eds.), Cambridge: MIT Press, pp. 521–544.

S. S. Rao and K. Chellapilla (1996), Evolving reduced parameter bilinear models for time series prediction using fast evolutionary programming, *Genetic Programming 1996: Proc. 1st Ann. Conf. on Genetic Programming*, J. R. Koza, D. E. Goldberg, D. B. Fogel, and R. L. Riolo (eds.), Cambridge: MIT Press, pp. 528–535.

J. Reed, R. Toombs, N. A. Barricelli (1967), Simulation of biological evolution and machine learning: I. Selection of self-reproducing numeric patterns by data processing machines, *J. Theoret. Biol.* 17: 319–342.

G. A. Riessen, G. J. Williams, and X. Yao (1997), PEPNet: Parallel evolutionary programming for constructing artificial neural networks, *Evolutionary Programming VI*, P. J. Angeline, R. G. Reynolds, J. R. McDonnell, and R. Eberhart (eds.), Berlin: Springer, pp. 35–46.

J. Risannen (1984), Universal coding, information, prediction and estimation, *IEEE Trans. Inf. Theory* 30: 629–636.

M. M. Rizki, L. A. Tamburino, and M. A. Zmuda (1993), Evolving multi-resolution feature detectors, *Proc. 2nd Ann. Conf. on Evolutionary Programming,* D. B. Fogel and W. Atmar (eds.), San Diego, CA: Evolutionary Programming Society, pp. 108–118.

M. M. Rizki, L. A Tamburino, and M. A. Zmuda (1995), Evolution of morphological recognition systems, *Evolutionary Programming IV: Proc. of 4th Ann. Conf. on Evolutionary Programming,* J. R. McDonnell, R. G. Reynolds, and D. B. Fogel (eds.), Cambridge: MIT Press, pp. 95–106.

R. Root (1970), An investigation of evolutionary programming, MS thesis, New Mexico State University, Las Cruces.

R. Rosenberg (1967), Simulation of genetic populations with biochemical properties, Doctoral dissertation, University of Michigan, Ann Arbor.

H. H. Rosenbrock (1960), An automatic method for finding the greatest or least value of a function, *Computer J.* 3: 175.

G. Rudolph (1997), Reflections on bandit problems and selection methods in uncertain environments, *Proc. 7th Int. Conf. on Genetic Algorithms,* T. Bäck (ed.), San Francisco: Morgan Kaufmann, pp. 166–173.

D. E. Rumelhart, G. E. Hinton, and R. J. Williams (1986), Learning internal representations by error propagation, *Parallel Distributed Processing,* vol. 1, D. E. Rumelhart and J. L. McClelland (eds.), Cambridge: MIT Press, pp. 318–362.

D. E. Rumelhart and J. L. McClelland (1986), PDP models and general issues in cognitive science, *Parallel Distributed Processing,* vol. 1, D. E. Rumelhart and J. L. McClelland (eds.), Cambridge: MIT Press, pp. 110–146.

R. Salomon (1996), Reevaluating genetic algorithm performance under coordinate rotation of benchmark functions: A survey of some theoretical and practical aspects of genetic algorithms, *BioSyst.* 39: 263–278.

N. Saravanan (1993), Evolving neural networks: Application to a prediction problem, *Proc. 2nd Ann. Conf. on Evolutionary Programming,* D. B. Fogel and W. Atmar (eds.), La Jolla, CA: Evolutionary Programming Society, pp. 72–78.

N. Saravanan and D. B. Fogel (1994), Learning strategy parameters in evolutionary programming: An empirical study, *Proc. 3rd Ann. Conf. on Evolutionary Programming,* A. V. Sebald and L. J. Fogel (eds.), River Edge, NJ: World Scientific, pp. 269–280.

N. Saravanan and D. B. Fogel (1995), Evolving neural control systems, *IEEE Expert,* June, pp. 23–27.

N. Saravanan and D. B. Fogel (1996), An empirical comparison of methods for correlated mutations under self-adaptation, *Evolutionary Programming V: Proc. 5th Ann. Conf. on Evolutionary Programming,* L. J. Fogel, P. J. Angeline, and T. Bäck (eds.), Cambridge: MIT Press, pp. 479–485.

N. Saravanan, D. B. Fogel, and K. M. Nelson (1995), A comparison of methods for self-adaptation in evolutionary algorithms, *BioSyst.* 36: 157–166.

N. Saravanan and D. B. Fogel (1997), Multi-operator evolutionary programming, *Evolutionary Programming VI: Proc. 6th Ann. Conf. on Evolutionary Programming,* P. J. Angeline, R. G. Reynolds, J. R. McDonnell, and R. Eberhart (eds.), Berlin: Springer, pp. 215–221.

J. D. Schaffer (ed.), (1989), *Proc. 3rd Int. Conf. on Genetic Algorithms,* San Mateo, CA: Morgan Kaufmann.

J. D. Schaffer, R. A. Caruana, L. J. Eshelman, and R. Das (1989), A study of control parameters affecting online performance of genetic algorithms for function optimization, *Proc.*

3rd Int. Conf. on Genetic Algorithms, J. D. Schaffer (ed.), San Mateo, CA: Morgan Kaufmann, pp. 51–60.

J. D. Schaffer and L. J. Eshelman (1991), On crossover as an evolutionarily viable strategy, *Proc. 4th Int. Conf. on Genetic Algorithms,* R. K. Belew and L. B. Booker (eds.), San Mateo, CA: Morgan Kaufmann, pp. 61–68.

A. V. Sebald (1991), personal communication, UCSD.

A. V. Sebald and K. Chellapilla (1998), On making problems evolutionarily friendly, Part 1: Evolving the most convenient representations, *Evolutionary Programming VII: Proc. 7th Ann. Conf. on Evolutionary Programming,* V. W. Porto, N. Saravanan, D. E. Waagen, and A. E. Eiben (eds.), Berlin: Springer, pp. 271–280.

A. V. Sebald and D. B. Fogel (1990), Design of SLAYR neural networks using evolutionary programming, *Proc. 24th Asilomar Conf. on Signals, Systems, and Computers,* R. R. Chen (ed.), San Jose, CA: Maple Press, pp. 1020–1024.

A. V. Sebald and D. B. Fogel (1992), Design of fault tolerant neural networks for pattern classification, *Proc. 1st Ann. Conf. on Evolutionary Programming,* D. B. Fogel and W. Atmar (eds.), La Jolla, CA: Evolutionary Programming Society, pp. 90–99.

A. V. Sebald and J. Schlenzig (1994), Minimax design of neural net controllers for highly uncertain plants, *IEEE Trans. Neural Net.* 5: 73–82.

A. V. Sebald, C. A. Sebald, and J. Schlenzig (1989), Use of neural net control strategies in difficult adaptive control problems, *Proc. 23rd Asilomar Conf. on Signals, Systems, and Computers,* vol. 1, R. R. Chen (ed.), San Jose, CA: Maple Press, pp. 342–345.

A. V. Sebald, J. Schlenzig, and D. B. Fogel (1991), Minimax design of CMAC encoded neural network controllers using evolutionary programming, *Proc. 25th Asilomar Conf. on Signals, Systems, and Computers,* vol. 1, R. R. Chen (ed.), San Jose, CA: Maple Press, pp. 551–555.

P. K. Simpson (1992), Fuzzy min-max neural networks: 1. Classification, *IEEE Trans. Neural Net.* 3: 776–786.

P. K. Simpson (1993), Fuzzy min-max neural networks: 2. Clustering, *IEEE Trans. Fuzzy Syst.* 1: 32–45.

M. Sipper, E. Sanchez, D. Mange, M. Tomassini, A. Pérez-Uribe, and A. Stauffer (1997), A phylogenetic, ontogenetic, and epigenetic view of bio-inspired hardware systems, *IEEE Trans. Evol. Comp.* 1: 83–97.

R. J. Solomonoff (1966), Some recent work in artificial intelligence, *Proc. IEEE* 54: 1687–1697.

M. Sternberg and R. G. Reynolds (1997), Using cultural algorithms to support re-engineering of rule-based expert systems in dynamic performance environments: A case study in fraud detection, *IEEE Trans. Evol. Comp.* 1: 225–243.

M. Tomita (1982), Dynamic construction of finite automata from examples using hill-climbing, *Proc. 4th Ann. Cognitive Science Conf.,* Ann Arbor, MI: Cognitive Science Sociey, pp. 105–108.

R. E. Trellue (1973a), Artificial evolution applied to character recognition, *PSL Tech. J.,* December.

R. E. Trellue (1973b), The recognition of handprinted characters through evolutionary programming, MS thesis, New Mexico State University, Las Cruces.

R. E. Trellue (1974), The recognition of characters using evolving automata, *Milwaukee Symp. on Automatic Control,* Milwaukee, WI, March 28–30.

A. M. Turing (1956), Can a machine think? *The World of Mathematics,* vol. 4, J. R. Newman (ed.), New York: Simon and Schuster, p. 2122.

R. W. Vincent (1976), Evolving automata used for recognition of digitized strings, MS thesis, New Mexico State University, Las Cruces.

J. von Neumann and O. Morgenstern (1947), *Theory of Games and Economic Behavior,* Princeton: Princeton University Press.

D. E. Waagen and J. R. McDonnell (1995), A combined stochastic and deterministic approach for classification using generalized mixture densities, *Evolutionary Programming IV: Proc. 4th Ann. Conf. on Evolutionary Programming,* J. R. McDonnell, R. G. Reynolds, and D. B. Fogel (eds.), Cambridge: MIT Press, pp. 159–174.

G. L. Williams (1977), Recognition of hand-printed numerals using evolving automata, MS thesis, New Mexico State University, Las Cruces.

B. R. Wolin (1963), Complex behavior in a "single task," *Hum. Factors* 5: 1–5.

D. H. Wolpert and W. G. Macready (1997), No free lunch theorems for optimization, *IEEE Trans. Evol. Comp.* 1: 67–82.

H.-S. Wong and L. Guan (1998), Adaptive regularization in image restoration using evolutionary programming, *Proc. 1998 IEEE Int. Conf. on Evolutionary Computation,* Piscataway, NJ: IEEE Press, pp. 159–164.

K. P. Wong and J. Yuryevich (1998), Evolutionary-programming-based algorithm for environmentally-constrained economic dispatch, *IEEE Trans. Power Syst.* 13: 301–306.

J.-M. Yang and J.-H. Kim (1997), Generation of optimal fault tolerant locomotion of the hexapod robot over rough terrain using evolutionary programming, *Proc. 1997 IEEE Int. Conf. on Evolutionary Computation,* Indianapolis, IN: IEEE Press, pp. 489–494.

X. Yao (1997), The importance of maintaining behavioural link between parents and offspring, *Proc. 1997 IEEE Int. Conf. on Evolutionary Computation,* Indianapolis, IN: IEEE Press, pp. 629–633.

X. Yao, G. Lin, and Y. Liu (1997), An analysis of evolutionary algorithms based on neighborhood and step sizes, *Evolutionary Programming VI: Proc. 6th Ann. Conf. on Evolutionary Programming,* P. J. Angeline, R. G. Reynolds, J. R. McDonnell, and R. Eberhart (eds.), Berlin: Springer, pp. 297–307.

X. Yao and Y. Liu (1996), Fast evolutionary programming, *Evolutionary Programming V: Proc. 5th Ann. Conf. on Evolutionary Programming,* L. J. Fogel, P. J. Angeline, and T. Bäck (eds.), Cambridge: MIT Press, pp. 451–460.

X. Yao and Y. Liu (1997), EPNet for chaotic time-series prediction, *Simulated Evolution and Learning,* X. Yao, J.-H. Kim, and T. Furuhashi (eds.), Berlin: Springer, pp. 146–156.

X. Yao and Y. Liu (1998a), Scaling up evolutionary programming algorithms, *Evolutionary Programming VII: Proc. 7th Ann. Conf. on Evolutionary Programming,* V. W. Porto, N. Saravanan, D. E. Waagen, and A. E. Eiben (eds.), Berlin: Springer, pp. 103–112.

X. Yao and Y. Liu (1998b), Making use of population information in evolutionary artificial neural networks, *IEEE Trans. on Syst., Man Cybern. Part B: Cybern.* 28: 417–425.

X. Yao and Y. Liu (1998c), Towards designing artificial neural networks by evolution, *Appl. Math. Computat.* 91: 83–90.

P. P. C. Yip and Y.-H. Pao (1994), Growing neural networks using guided evolutionary simulated annealing, *Proc. 3rd Ann. Conf. on Evolutionary Programming,* A. V. Sebald and L. J. Fogel (eds.), Singapore: World Scientific, pp. 17–25.

INDEX

Ackley's function, 110
Adaptation, 14, 135, 142
Adaptive landscape (also fitness landscape), 121, 138
Adaptive systems, 80
Adversary, intelligently interactive, 133
Akaike's information criterion (AIC), 88, 89, 115
Annealing, 72
 simulated, 99, 114
Apple II, 7
Armed conflict, 133
Artificial intelligence, *see* Intelligence, artificial
Artificial life, 86, 142
Australian Defence Force Academy, 118
Automata, 2
Automobile insurance claims, 96
Autonomous vehicles, 80, 82, 83

Backpropagation, 114
Barricelli, Nils, 46, 142
Behavior, 35, 56, 91
 conscious, 35
 cooperative, 127
 learned, 65
 social, 43
Behavioral response, 80
Bienert, Peter, 46
Bit (binary) strings, 137–139
Bledsoe, Woody, 46

Bohachvesky function, 100, 101, 112
Bomber, 56
Box, George, 142
Brain, 3
Bremermann, Hans, 46, 142
Broom balancing, 121
Building blocks, 80, 121

Characteristic cycle, 20
Circuit design, 136
Classification, 28, 44, 60, 62, 91, 113. *See also* Pattern classification
 breast cancer, 118
 spiral, 121
Clustering, 74, 91–93, 96–98
Coevolution, 50, 52, 56, 86, 115
Combat, 50, 133
Commodore 64, 7
Communication, 35
Community, 38
Competition, 65
 round-robin, 94
Complexity, 44
Computer:
 digital, 40
 -generated forces, 133
 speed, 31
Conrad, Michael, 46, 142
Constrained optimization, 136
Constraints, 140

Convergence properties, 85, 136
Control, 28, 34, 35, 44, 57, 68, 135
   classic theory, 28
   of dynamic systems, 121
   evolutionary system, 29
   fuzzy, 135
   system design, 28
   of systems with friction, 135
Convergence, 32
   asymptotic, 63, 64, 136
   on a goal, 38
   logarithmic, 136
   rate, 53, 139
Cooperate, 123, 127
   mutual (cooperation), 124–126
Creativity, 39, 45, 139
Critical, 140
   degrees of criticality, 141
Crossover, 72, 74, 80, 84, 85, 100–102, 117, 118, 119, 138, 139
   headless chicken, 121
   partially mapped (PMX), 75, 76
Cultural algorithms, 96
Cutting stock optimization, 135

Data mining, 96
Data structures, 46
Dearholt, Don, 32, 46, 59, 65
Decision-making, 1, 2, 32, 33, 35, 36, 38–42, 44
Decision Science, Inc., 46
Deductive manipulation, 37
Defect, 123, 127
   mutual (defection), 124
Detection, 20, 21, 28
Discrimination, 21, 28

Earthquake location, 135
Ecosystems, 65, 66
Efficiency, 16
   of evolution, 31, 35
Electrocardiograms, 62, 64
Electronic part placement, 135
Email, 142
Environment, 3, 4, 6–9, 12–14, 16, 18, 33–39, 42–45, 54, 55, 73
   antagonistic, 38
   artificial, 65
   battlefield, 133
   binary, 68
   changing, 73
   complex, 31
   conditional properties of, 21
   cyclic, 8–10, 12–15, 68, 72
   dynamic, 71

   logic of, 34
   noisy, 10, 34, 55, 139
   nonstationary, 10, 16, 18, 31
   obverse, 72
   periodic, 20
   radical change in, 10, 14
   repetitive, 21
   shift, 15
   stationary, 14, 15
   stochastic properties of, 20
   two-symbol, 7
Equilibrium point, 51, 52
Evaluation criteria, 28
Event, rare, 18
Evolution, 3, 22, 36, 39, 42, 58, 94, 139, 140
   artificial, 39
   asynchronous, 140
   Darwinian, 140
   goals and, 42
   Lamarckian, 140
   nonregressive, 29
   simulating, 44, 46, 138
Evolutionary algorithm, 18, 40, 75, 85, 122, 136–139
Evolutionary computation, 46, 80, 119, 136, 137, 139–142
Evolutionary optimization, 63, 80, 121, 142
Evolutionary process, 38, 44
Evolutionary program, 17, 19, 21, 24, 27, 29, 41, 49, 50, 53, 55, 56, 60, 62, 71, 75, 76, 105, 115, 131, 134, 135
Evolutionary programming, 3–7, 18, 19, 30–35, 39, 40, 46–50, 52, 57, 59–62, 65, 68, 71–76, 80, 85–87, 92, 96, 100–105, 110, 111, 114, 115, 118, 119, 121–123, 125, 128, 133, 135–138
   meta-, 100–103
   theory, 136
Evolutionary simulations, 80
Evolution strategies, 46, 104, 136, 138
Exemplars, 87
Expert systems, 3
Exploration, 38

Factory scheduling, 135
Family, 38
Fibonacci series, 68
Finite element problems, 136
Finite state machine(s), 4–12, 16, 17, 19–21, 27, 29–31, 39, 41, 44, 47, 48, 50, 52, 53, 55–60, 64, 65, 68, 74, 111, 123–125, 127, 133
   complexity, 22
   self-adaptation of, 105

First Annual Conference on Evolutionary
    Programming, 104
Fogel, David, 7, 59, 68
Fraser, Alex, 46, 142
Friedberg, Richard, 142
Friedman, George, 142
Fuel distribution, 135
Function(s):
    autoregressive, 114
    continuous, 96, 104, 137
    convex, 136
    linearly separable, 136, 137
    optimization, 46
Fuzzy membership functions, 86

Galileo, 142
Game(s) (also game playing, gaming), 33, 46, 49,
        52, 122, 133
    bargaining, 52
    coordination, 50–52
    minimax value, 49
    nonzero-sum, 49–51, 122
    object of the, 131
    perfect information, 53
    rules, 53
    theory, 33, 46
    tree, 130, 132
    trust, 50–52
    two-player, 48, 51
    zero-sum, 47, 49, 50
Generalized mixture densities, 136
Generation, 7–10, 12–15, 17, 21, 72, 73, 80, 89,
        97, 98, 101, 125, 131, 140
Genetic algorithms, 31, 72, 100, 102–105, 119,
        121, 123, 136–138
    premature stagnation in, 104
Goal, 34, 35, 38–40, 42, 43, 45, 134
    seeking ability, 34, 35, 39, 42
    subgoal, 43
Gradient search, 88

Handwritten characters, 59, 60, 62, 65
Hardware, realization of evolutionary machines
        in, 65, 140
Heredity of reasonableness, 39
Heuristic programming, 3
Heuristics, 39, 133
Hill climbing, 121
Holland, John, 46, 142
Human operator, 56–59

IBM, 7
Ice cracking, 87–91
Identification, 68

IEEE Neural Networks Council, 142
Image analysis, 135
Image restoration, 136
Imagination, 39, 45
Individual, 38
Induction, 36, 39
    language, 117
    of sequences of symbols, 68
Inductive inference, 37, 39
Information processing, 40, 42
Inheritance, 39
Input-output response, 57, 113
Intellect, 2, 3, 21, 39, 43, 45
    upper bound, 40
Intelligence, 2, 3, 39, 41–44, 65
    artificial, 3, 21, 28, 44, 85, 86, 142
    forms of, 65
    level of, 40
    logical limitations of, 41
    nature of, 65
    ontogenetic, 65
    phylogenetic, 65
    sociogenetic, 65
Intelligent, 42, 44
    adversary, 35, 42
    behavior 1–3, 44, 133
Interceptor, 56
Internet, 142
Inversion, 80, 84
Iris data (Fisher), 91, 94

k-armed bandit, 137
Knowledge, 2, 39
    incomplete, 28
Knowledge-based systems, 3

Learning, 18
    curves, 63
    improved initial, 15
    individual, 2
    rate, 13, 60, 64, 129, 131
Ligand docking, 135
Linear equations (systems of), 80, 84, 86
Localizing ships, 135
Local optima, 114, 138

Majority logic, 31–33
Markov sequences (transitions), 68
    first-order, 19, 20
    zero-order, 19, 20
Mealy machine, 4
Memory, 34
    size of, 40
Migration, 65

160 INDEX

Minimax:
  players, 46
  solution, 50
Minimum description length principle (MDL), 91–94, 97, 98
Model, 34, 37, 38, 40, 42, 44, 45, 57, 87, 89, 91, 94
  autoregressive moving-average (ARMA), 87, 88, 92
  first-level, 38
  high-level, 38
  parameters, 87
    of flight data of the X-15, 59
    of itself, 35
    of models, 37
  order, 87
ModSAF, 133
Moore machine, 59, 60
Multiobjective optimization, 141
Multiple interacting programs (MIPs), 119, 121, 122
Mutation(s), 4, 6–10, 14, 15, 17, 29, 38, 44, 65, 72–75, 80, 84, 85, 120, 128, 138
  adaptable probabilites of, 60, 86, 96
  based on evolutionary process, 31
  bit, 100
  Cauchy, 107, 108, 110
  component level, 105
  convolution, 109
  correlation between, 104
  degree of severity, 107
  distribution, 105
  Gaussian, 84, 100, 104–106, 108–110, 113
  individual level, 105
  lognormal, 104–106, 109
  macro-, 121
  multiplicity of, 10, 12, 13, 41, 60, 75, 107
  noise, 8, 31
  parameters, 100, 105
  per parent, 12, 71
  Poisson distributed, 88
  probability of adding a state, 9, 12, 31
  rates, 60, 71
  self-adaptation, 96, 104, 105, 109, 138, 139
    as histograms, 109, 112
  strategies, 100
  variance, 100, 113
  wild noise, 19

Natural evolution, 2
Natural selection, 2
Naughts and crosses, *see* Tic-tac-toe
Neural networks, 2, 3, 86, 113, 114, 118, 121, 127–130
  associative memory, 118
  evolving, 111, 119
  for control, 114
  Hopfield, 118
  modular, 119
  optimizing architectures, 117
  recurrent, 115
  threshold, 117
Newell, Alan, 131
New Mexico State University, 32, 46, 59, 65
Noise, 10, 21, 34
  Gaussian, 21, 22, 100
Nonminimax players, 46, 47

Offspring, 4, 6–8, 29, 31, 39, 40, 44, 60, 65, 74, 75, 83, 85, 86, 88, 89, 92, 100, 108, 113, 117–119, 136, 138
  variable number of, 60

Parallel data processing, 41
Parent(s), 4, 6, 8, 19, 29, 31, 60, 65, 74, 75, 80, 86, 89, 92, 100, 113, 117–119, 125, 128, 138, 140
  machine, 8, 10, 44, 50, 72
  variable number of, 60, 136
Pareto optimal set, 141
Pattern, 20, 22, 23, 27, 87, 91
  classification, 21
  as compared with humans, 21
  recognition, 21, 46, 59, 65, 111
Payoff, 122, 123
  all-or-none, 16
  function (matrix), 3, 5, 16, 20, 21, 29, 34, 47, 48, 50, 52, 53, 71, 72, 128–130, 132, 141
  optimal expected, 48
  sucker's, 123
  temptation, 123
Penalty, 82
  for complexity, 9, 10, 14, 15, 17, 19, 21, 50, 60, 72
  term, 93, 94
Plant, 28, 29, 34
  displacement, 58
  linear, 28
  second-order, 57, 58
  steady-state response, 28
  transient response, 28
Platoons, 135
Polar coordinates, 105
Population, 3, 6, 14, 29, 47, 50, 60, 65, 68, 73, 80, 84, 85, 92, 98–100, 114, 115, 121, 124, 125, 136, 137, 139
Porter, Cole, 142

Prediction(s), 5–9, 11, 14, 17, 18, 20, 27, 44, 53, 54, 58, 61, 65, 68, 71–74, 96, 118, 122
  as compared with humans, 18, 19
  sunspot, 121
  water temperature of cooling tower, 59
Predictive logics, 18
Predictors, 12, 15, 20
Prime numbers, 16–18, 61
Prisoner's dilemma, 51, 52, 122, 124
  continuum of behavior, 127
  duration length parameter, 125, 126
  iterated, 20, 49, 122, 125, 127
Problem solving, evolutionary, 40
Protein-protein interactions, 135
Pursuit-evasion, 52–56, 86, 135

Query translation, 135

Random, 20
  disturbance, 34
  search (blind), 71, 118
  variable (Gaussian), 92, 118
  variation, 139
Randomness, 19, 39
Rate of optimization, 79, 81–83, 94, 96–98, 114
Real-world, 34, 137
Recall, 53
  length, 13, 27, 50
  length of initial, 10
  period, 15
Rechenberg, Ingo, 46, 142
Recognition, 28, 45
Re-cognition, 21
Recombination, 2, 31, 72, 75, 80, 85, 117, 138
  intermediate, 136
  of subtrees, 119, 121
Recursive prediction error method (RPEM), 89, 91, 92
Regular expression, 59, 60
Regularity, 20
Representation, 85, 86, 138
  based on minimum cardinality, 137
  order-based, 74
  parse tree, 119–121
  problem-specific, 86
  real-valued, 137
  variable-length, 138
Response, 34
  conditioned, 35
  surface, 121, 140
Rosenbrock function, 100, 102, 111
Routing problems, 75
Royal road problem, 121

Rule-based, 134
  behaviors, 135

Saddlepoint, 51
Schema processing, 137, 139
Schwefel, Hans-Paul, 46, 142
Scientific method, 36–39, 44, 45
Selection, 4, 6, 7, 14, 38, 41, 44, 75, 100, 119, 138–140
  probabilistic, 32, 75, 80, 86, 138
  proportional, 32, 137–139
  roulette wheel, 31
Self-adaptive, 60, 86, 88, 100, 102, 113
Self-awareness, 35, 42, 45, 140
Self-preservation, 42, 45
Self-reference, 34
Sequences, 74
  complex, 19
  cyclic, 72
  statistical, 19
Signal, 21
Simulation, 133
Squared error, 88
Social groups, 38, 45
Sonar returns, 113
Sound, sources of, 87
Species, 38
Stimulus-response, 35, 36, 123
Stochastic:
  disturbance, 28
  process, 21
Strategies, 122, 133, 140
  evolved, 53
  optimal, 33
  time-varying, 49
Survival, 65
  probabilistic, 94
  probability of, 135
Symbolic (S-)expressions, 86, 119
System:
  linear (dynamic), 56, 57
  nonlinear (dynamic), 56, 58, 59

T-C problem, 115, 116
Technical University of Berlin, 46
Tic-tac-toe, 127, 129–132
Time series, 56, 59, 87, 91, 96, 118, 122
Tournament, 89
  round-robin, 129
Training, 114
Traveling salesman problem, 72, 74, 76, 79, 86, 105, 107
  2–OPT, 76
  inversion, 77

Traveling salesman problem *(continued)*
   multiple, 74, 80, 81
   three-dimensional, 74, 75
   two-person, 74
   uniform, 74, 75
Turing, Alan, 39

University of California, Berkeley, 46
University of Michigan, 46

Valuated State Space, 141
Variation, 140

operators, 75, 76, 85, 86, 121, 139
random, 39, 115, 119
step size of, 10
Virtual world, 133
VLSI design, 135
Von Neumann, John, 33, 47, 142

World Congress on Computational Intelligence, 142

Yao, Xin, 118